CATALYTIC PROCESSES AND PROVEN CATALYSTS

CATALYTIC PROCESSES AND PROVEN CATALYSTS

CHARLES L. THOMAS

Tempe, Arizona

Retired Scientific Advisor
Research and Engineering Department
Sun Oil Company

1970

ACADEMIC PRESS New York and London

ACADEMIC PRESS, INC.
111 Fifth Avenue, New York, New York 10003

United Kingdom Edition published by
ACADEMIC PRESS, INC. (LONDON) LTD.
Berkeley Square House, London W1X 6BA

LIBRARY OF CONGRESS CATALOG CARD NUMBER: 78-116700

PRINTED IN THE UNITED STATES OF AMERICA

CONTENTS

7. Catalytic Reforming of Naphthas

8. Polymerization

9. Alkylation

10. Catalysts for Free Radical Polymerizations

11. Synthesis Gas and Hydrogen Production

12. Gas Purification

13. Hydrogenation

14. Synthesis of Ammonia and Methanol

15. Hydrogenation of Carbonyl Groups

16. Hydrorefining

17. Hydrocracking of Petroleum and Coal

18. Oxidation (Inorganic)

19. Oxidation (Organic) Liquid Phase

20. Oxidation (Organic) Vapor Phase

21. Miscellaneous Processes

22. Catalyst Carriers and Supports

PREFACE

The objective of this book is to help the practicing chemist or chemical engineer, academic, governmental or industrial, who has a problem he thinks might be solved by means of a catalyst. It does this by presenting the catalysts available through commercial U. S. suppliers and that are known to be active in some process, usually a commercial one. The presentation is via the process in which the catalyst is used giving an outline of the conditions under which the catalyst has been found useful. Hopefully, this will provide the reader having the problem some ideas of proven catalysts that he can apply to his project or at least some ideas of the types of materials he should try as catalysts.

The reader is certainly aware that industrial users of catalysts do not ordinarily give the details of their commercial operations. So the use conditions given must be regarded as either approximations or obsolete with the best conditions closely guarded by the user.

This book is not intended as a scholarly work so there is not a complete literature summary for each topic. Rather, the author has given references that should give the reader a "lead" into the literature in case he wants more information. The author offers his apologies to those authors who may feel slighted because their work is not mentioned or for work that is mentioned that they regard as of less significance than their own.

The author emphasizes that the purpose is to help the reader solve his problem—not to solve it for him. As a result, few readers will find all the information in it that they would like to have. Readers who are already expert in the field of catalysis will find little in it that they are not already aware of and will doubtless have more advanced or more precise information in the fields in which they are specialists.

The original version of this book was prepared by the author as a private edition for use inside Sun Oil Company. It is a pleasure to record my gratitude to Mr. C. G. Kirkbride, Vice President for Research and Engineering, for encouraging me to prepare this version for publication and for permission to use material from the Sun Oil Company edition.

It is also a pleasure to record my gratitude to the many catalyst suppliers who reviewed their portions of the Sun Oil Company edition, suggested improvements and granted permission to include the data in this book.

Tempe, Arizona
April, 1970

CHARLES L. THOMAS

CATALYTIC PROCESSES AND PROVEN CATALYSTS

CHAPTER 1

USING CATALYSTS

I. Catalysts Are Chemicals

By keeping one simple fact in mind the catalyst user can save both time and grief: a catalyst is a chemical and the substance being catalyzed forms an intermediate chemical compound with the catalyst. This intermediate reacts to give the product or with a second reactant to give the product. To fulfill its catalytic nature, the intermediate must have an optimum stability. If it is too stable, it will not react fast enough to give a usable reaction rate. If it is not stable enough, it will not be formed either fast enough or in sufficient concentration to provide a useful reaction rate. The intermediate must both form and react (decompose) at useable rates.

It is apparent that any solvent used in the reaction must not form a stable compound with the catalyst or the catalyst will be poisoned. Similarly, impurities which form stable compounds with the catalyst must not be present in the solvent or reactant.

These considerations apply whether the catalyst is homogeneous or heterogeneous, liquid or solid. Most of the catalysts treated in this book are solids and the following discussion applies to them.

II. Reactant Transport

Whether liquid or vapor, a reactant must get to the catalyst before it can react with the catalyst. Likewise, the product must get away from the catalyst before more reactant will be able to react. Both of these are physical processes which to some extent can be controlled by external forces, such as rate of stirring, flow rates, etc., which will be discussed in more detail in specific operations.

The catalytic process can then be summarized by the following steps:

(*A*) Diffusion of the reactant to the catalyst,
(*B*) Reaction with the catalyst (chemisorption),
(*C*) Reaction of the intermediate to give the product,
(*D*) Desorption of the product,
(*E*) Diffusion of the product away from the catalyst.

Not all of these steps are always present, and rarely is it possible to follow them individually. Even so, the catalyst user must be aware that any of the steps may be rate controlling. For example, if (*A*) or (*E*) is rate controlling, the energy of activation will be <5 kcal mole^{-1}. Such an energy of activation suggests that the catalyst is not being used with maximum effectiveness and that effectiveness can be improved by improving mass transfer conditions.

Steps (*B*), (*C*), and (*D*) being chemical reactions, they will have energies of activation above about 10 kcal mole^{-1} and thus no further improvement can be expected by improving mass transfer unless heat requirements (endo- or exothermic) are involved.

There are several major types of reactors for using catalysts.

III. The Stirred Reactor

This is the most common laboratory technique and is one widely used in industry. It ordinarily utilizes a finely divided solid catalyst suspended in a liquid contained in a vessel so the liquid can be agitated. The vessel may be a laboratory flask plus stirrer, an autoclave, or a large industrial kettle.

The reactor is used by mixing a few percent (0.1–5%) catalyst with the liquid, bringing the mixture to the temperature and pressure of reaction conditions, and holding it there until the reaction is complete. Used in this way, the operation has the advantage of being able to take the reaction to completion or to stop it at a partial conversion, as desired. The catalyst may be recovered for reuse, but it loses activity as the number of subsequent batches increases. Therefore, a portion of the catalyst is removed and replaced with fresh catalyst to maintain a constant activity. Alternatively only fresh catalyst is used each time.

The stirred reactor need not be used as a batch reactor for the product may be withdrawn intermittently or continuously and replaced by reactant (fresh feed). This latter use inherently means that the product will always be mixed with reactant, i.e., the reaction will be only partly complete. If the two materials are readily separated, this is no problem. Where greater conversion of the reactant is desired, two or three stirred, continuous reactors are used in series and the product withdrawn from the last reactor. Whether used singly or in series, the retention time in each reactor is an important variable, along with temperature, pressure, and catalyst concentration, in determining the degree to which the reaction goes to completion.

With either batch or continuously stirred reactors, heat evolution during reaction is often a problem. Conducting the reaction under conditions such that the liquid (reactant or solvent) can be refluxed is one stratagem used to dissipate the heat and control the temperature.

IV. The Fixed Bed Reactor

This apparatus is also common in both the laboratory and industry. In its simplest form it comprises a column or "bed" of catalyst granules in a confined space and a liquid, gas, or vapor passed through the bed—usually downward. The catalyst container (catalyst "case") in the laboratory is surrounded with some device for controlling the temperature, usually a heater. In the plant the reactor is most often adiabatic, but in a few processes it is heated or cooled.

A technique widely used in the laboratory involves putting a fraction of a gram of catalyst in a microreactor and passing the reactant through the catalyst with a carrier gas, usually H_2 or He (*1*). The effluent is taken directly to a vapor chromatograph. Since the reactant is introduced in pulses, the technique is referred to as the "micropulse" method. The method has gone through a number of evolutionary changes, including the addition of a convenient pressure unit that operates at pressures up to 70 atm (*2*).

When an appreciable amount of product is wanted for use, a larger reactor is employed and separate arrangements are made for product analysis.

With the fixed bed reactor, the reactant can be passed through the catalyst fast or slowly. The relative rate is usually referred to as the "space velocity." This is defined as the volume of reactant per gross volume of catalyst bed per hour and is expressed as LHSV, the liquid hourly space velocity, i.e., $V_l V_c^{-1}$ hr^{-1} if the reactant is a liquid. If the reactant is a gas, the GHSV is the gas hourly space velocity, i.e., $V_g V_c^{-1}$ hr^{-1}. V_l or V_g, the volume of the gas or liquid, is measured at room temperature and atmospheric pressure. It is amazing how useful such a loose term has been in catalysis.

A somewhat more precise term is the weight hourly space velocity, WHSV, i.e., $W_l W_c^{-1}$ hr^{-1}. It is related to $V_l V_c^{-1}$ hr^{-1} as follows: $V_l d_l V_c^{-1} d_c^{-1}$ hr^{-1} where d_l is the density at room temperature of the liquid and d_c is the *apparent bulk density* (ABD) of the catalyst granules.

The LHSV for fixed bed reactors commonly falls in the range of 0.5–10, with most in the narrower range of 1–4 $V_l V_c^{-1}$ hr^{-1}. This is a convenient rate in the laboratory and in the plant it usually permits the ready handling of heat and pressure drop problems.

Several process terms besides temperature, pressure, and space velocity need defining for fixed bed reactors. The term "time on stream" or "process period" refers to the length of time the reactant is passed over the catalyst bed without interruption, especially, for regeneration. So far as the author knows, all catalysts lose activity with use. The loss may be fast, i.e., in a matter of minutes (as in the catalytic cracking of petroleum) or may be slow, i.e., in years (as with iron synthetic ammonia catalysts).

"Conversion per pass" refers to the percent or fraction of the feed which disappeared (was converted) in one pass through the catalyst. As space velocity increases, conversion per pass decreases. $(SV)^{-1}$ is frequently used as a correlating variable instead of SV, because at $(SV)^{-1} = 0$, conversion = zero, i.e., an extra point on the curve is given. For first-order or pseudo first-order reactions, the following relation is useful:

$$K = -SV \log \frac{X}{X_0} \tag{1}$$

in which K is a constant related to the first order reaction velocity constant, X_0 is the initial concentration of reactant and X is the fraction of the reactant unconverted under the conditions used. A plot of log X versus $(SV)^{-1}$ gives a straight line for this type of reaction. Changes in K with temperature can be used to determine the energy of activation for the reaction via the Arrhenius equation.

Equation (1) may also be used to assign a number to the relative activity of a series of catalysts. The catalyst that gives the smaller value of X at constant test conditions is the more active catalyst. Equation (1) then permits calculating the space velocities for the series of catalysts to give constant X. These space velocities are more valid expressions of relative activity than X itself. If WHSV is used, then catalyst density effects such as those illustrated in Fig. 1 below can be brought into consideration. For commercial operations, both volume and weight space velocity values can be useful because catalyst is purchased by weight and used in fixed bed reactors by volume, i.e., the volume of the commercial reactor is fixed. A more active catalyst may permit an increase in plant capacity with relatively little cost.

It must be kept in mind that it is very common for fresh (unused) catalysts to have a highly active "edge" of activity which is rapidly worn off in use after which the catalyst settles down to a lower but more stable level of activity. Measurements made while the "edge" is wearing off all too often lead to confusion rather than useful information.

A. Catalyst Granule Porosity

When the catalyst is in a finely divided state, such as that used as a suspension in the stirred reactor, the reactant has relatively little difficulty in getting to all the surface of the catalyst. However, in the fixed bed reactor with catalyst granules, some of the active surface is inside the granules. If this surface is to be useful, the granule must be porous enough for the reactant to penetrate the particle and for the product to escape.

From a practical viewpoint, a relatively hard durable particle is desired, i.e., one that does not attrit to form fines that increase pressure drop across the catalyst bed. However, physical durability must be compromised with porosity that tends to make weak particles or granules.

The compromise can be illustrated by the following tests. A single batch of finely divided catalyst was separated into several aliquots. Each aliquot was pressed into granules about 2 × 2 × 2 mm but with varying pressure, such that the lightest pressure gave granules that were highly porous and would just hold together. With each succeeding aliquot, the pressure was increased so that the granules got progressively stronger and less porous. Granule density and apparent bulk density of the granules increased as the pressure increased.

The activity of each catalyst was tested at constant $V_l V_c^{-1}$ hr^{-1} using the same *volume* of catalyst granules in each test with the results shown in Fig. 1.

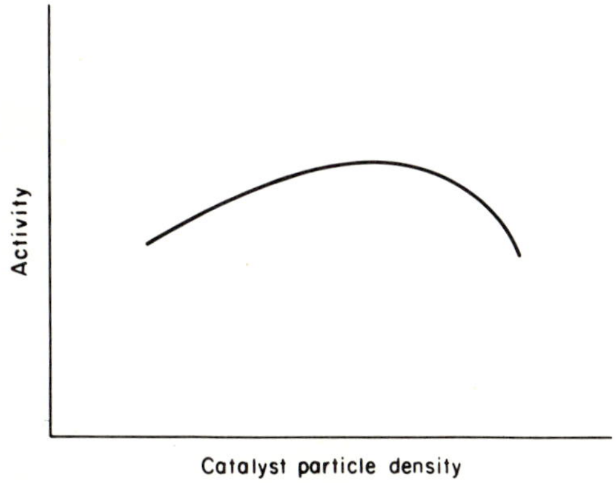

Fig. 1. Density affects apparent catalyst activity.

The apparent activity of the catalyst increased at first as the compressing pressure was increased because a greater weight of catalyst (more active surface) was in the reactor. As the catalyst was pressed still harder, more active surface went into the reactor; however, higher pressure on the granules crushed many of the pores, so the additional active surface was not available. As a result, the activity reached a maximum and further pressure actually reduced the apparent activity. Both the user and the catalyst manufacturer benefit when the catalyst porosity (apparent bulk density or particle density) is adjusted for optimum activity.

In describing their catalysts, some catalyst manufacturers supply the ABD (apparent bulk density), surface area, and pore volume. If available to the author, these are given as the various catalysts are described later in this book. A pore volume of about 0.3 $cm^3\ gm^{-1}$ or larger indicates a reasonably open structure in the granule.

Granule size is also related to the availability of active surface inside the granule. Small granules are preferred if the pressure drop can be tolerated across a bed made up of them.

The method of forming granules also has an effect on the availability of the internal surface. Very often granules have a "skin" that is less porous than the body and this inhibits flow to the inside. This is especially likely with pills or tablets made on a tabletting machine, although extrusions sometimes have a "skin" too.

B. Mass Velocity

A fixed bed reactor at constant space velocity can be long and narrow or short and wide. The short wide reactor is most often used when a low pressure drop across the bed is needed. The mass velocity (weight of reactant passing a unit cross section of the reactor in unit time) in this reactor is low.

The long narrow reactor (such as a tube or pipe) is used when heat effects in the reaction need careful control, i.e., when the reaction is strongly exothermic or endothermic, and this heat must pass through the walls of the reactor. A high mass velocity favors this heat transfer, but the pressure drop is large.

A large mass velocity can be used when the reaction tends to be diffusion controlled since the mass velocity also improves diffusion to and from the catalyst particle.

V. The Trickle Reactor

When the fixed catalyst bed is used with a liquid or mixed liquid plus vapor reactant under operating conditions, the trickle reactor is used, especially if the liquid is to react with a gas, e.g., in a hydrogenation. With this reactor, the liquid reactant is fed in at the top of the reactor so that it is distributed over the area of the catalyst bed. Each particle of catalyst is wet by the liquid as it "trickles" down through the bed. The reacting gas penetrates the liquid film on the surface of the catalyst and reacts. The catalyst bed may contain two or three redistributor plates along the direction of flow to prevent channeling of the liquid. The reacting gas may be recirculated (preferred) or a pressure maintained to replace the gas as it reacts without recirculation.

VI. The Jiggling Catalyst Bed

If a gas or vapor is passed upward through a fixed catalyst bed, it tends to lift the bed. If the velocity is great enough, the catalyst will be blown out of the reactor. At intermediate velocities the catalyst granules will rise, fall, and move about, i.e., "jiggle." This type of operation has several useful characteristics. Part of the catalyst can be removed and fresh catalyst added without interrupting the process. Also, the exothermic heat of reaction is distributed throughout the bed, rather than localized as it is in the fixed bed. This, in turn, permits relatively cool reactants to be introduced into the reactor so that the heat of reaction is used to supply needed preheat.

VII. The Moving Bed of Catalyst

When catalysts rapidly lose activity, it is difficult to use a fixed bed in a continuous process because the product yield is declining continuously as the catalyst ages. By moving catalyst granules through the reactor, it is possible to introduce fresh catalyst continuously and so maintain a constant catalyst activity. This permits a continuous operation with a constant product yield. The average catalyst activity is a function of the rate at which catalyst is fed to the reactor compared to the rate of reactant, i.e., the catalyst:reactant ratio ($W_cW_r^{-1}$) expressed as a weight ratio. The larger this ratio, the greater the average catalyst activity.

Three process variables have been introduced above to define the conditions under which catalysts operate:

Space velocity (SV)
Process period (C_t)
Catalyst to reactant ratio ($W_cW_r^{-1}$)

For the moving bed, the process period is the length of time the catalyst remains in the reactor. For the WHSV, the three are related as follows:

$$\text{WHSV} = \frac{1}{W_c W_r^{-1}} (C_t)^{-1}$$

for C_t expressed in hours.

It is customary to use the catalyst:reactant ratio as a primary operating variable and to express the "severity" of operating conditions by listing the catalyst residence time and the catalyst:reactant ratio.

In this type of operation, facilities for catalyst regeneration are located adjacently, so the regenerated catalyst can be recycled to the reactor immediately.

This system is rarely used in the laboratory.

VIII. The Fluidized Bed

If a layer of powdered catalyst is placed on the sintered disc of a filter funnel and air passed upward through the catalyst, a velocity will be found at which the bed expands and moves around, i.e., it behaves very much like a liquid or fluid. This is one way of contacting a gas or vapor with a catalyst. It has the advantages of the "jiggling" bed described above plus the following:

(*A*) Since a powdered catalyst is used, less attention need be paid to granule porosity.

(*B*) Particle attrition is less of a problem.

(*C*) As a pseudo liquid, the catalyst is easy to move in and out of the reactor or regenerator.

The system has the disadvantage that the catalyst powder tends to be carried out of the reactor and must be trapped in some way. In the laboratory, this is usually accomplished by using sintered discs or thimbles. In the plant, it is done with cyclone separators and electric precipitators.

The process variables of the moving bed catalyst system are equally applicable to the fluidized catalyst system in which catalyst is continuously removed and replaced by fresh or regenerated catalyst.

Most of the units of this type are used with catalyst that deactivates rapidly, and the catalyst is removed continuously, regenerated, and returned to the reactor. However, in some units the catalyst stays in the reactor all the time as a fluidized fixed bed. Here the fluidized bed is used in reactions that are strongly exothermic, e.g., oxidations. They are useful because the rapidly moving catalyst dissipates local hot spots (the temperature throughout a large fluid catalyst bed will usually be constant to $\pm 2°C$). Also, the heat of reaction can be removed, just as from a liquid, by putting catalyst coolers in the reactor.

This technique is used to a limited extent in the laboratory but, commercially, the largest catalytic units in the world are of this type, the units for the catalytic cracking of petroleum.

IX. Conversion versus Selectivity

In many catalytic reactions, the primary or initial product formed is capable of reacting further to give secondary reaction products. As one attempts to maximize conversion to the initial product, more and more opportunity is afforded this initial product to react further. The usual procedure limits the initial conversion so that the minimum secondary product is formed, separates the initial product, and recycles the unconverted material along with more reactant. There are no general rules that permit a prediction, but usually the secondary product will appear when 20–50% conversion has taken place.

X. Naming Catalysts

Rarely is the exact chemical identity of a catalyst known under reaction conditions. Even so, it is convenient to have names for catalysts that give some clue to composition, although they may be imprecise or even inaccurate. Names such as chromia–alumina or silica–alumina give some idea of their composition and are commonly used, although imprecise. Names such as "cobalt molybdate" for a catalyst that starts as a cobalt oxide–molybdenum oxide–aluminum oxide composition are used and do give some idea of what went into the catalyst; but they are inaccurate because the catalyst is not cobalt molybdate and the active form is some sulfide. Similar names are "copper chromite" and "zinc chromite." It is unfortunate that the neophyte in catalysis must get accustomed to such nomenclature, but better precision is lacking.

XI. Getting Catalyst Samples

Most manufacturers are more than glad to furnish *gratis* experimental samples of their catalysts for tests or research purposes (noble metal catalysts excepted). A few suppliers with proprietary interests to protect will decline. Whether the reader is in academic, governmental, or industrial research, he is urged to make his request for catalysts through the purchasing department of his own organization, making clear exactly what he wants. Many of the catalyst suppliers have brochures on their catalysts and are glad to furnish them. These contain useful technical information about the specific catalyst and its use. These brochures should also be requested through the reader's own purchasing department.

REFERENCES

1. Kokes, R. J., Tobin, H., Jr., and Emmett, P. H., *J. Am. Chem. Soc.* **77,** 5860 (1955).
2. Steingaszner, P., and Pines, H., *J. Catalysis* **5,** 356 (1966).

CHAPTER 2

BURNING COKE OFF CATALYSTS

I. Source of Coke

Many catalysts which are used to catalyze organic reactants have coke deposited on them and lose activity because of the coke. It is often economic to remove this coke by burning and then reuse the catalyst. Although sometimes referred to as "carbon," the catalyst deposit is rarely carbon but contains hydrogen and often oxygen, sulfur, and even nitrogen compounds.

II. Burning in a Fixed Bed

If one passes air through a coked fixed catalyst bed hot enough to get ignition (about 500°C), burning starts as the air hits the catalyst and forms a "hot spot" or hot layer. This moves through the bed as a discrete layer which can be readily followed by an axial thermocouple. If air is furnished fast enough, high temperatures are reached which damage the catalyst. The author has seen a hole melted in the wall of a steel tube from uncontrolled regeneration by operators not familiar with the hot spot. A moving thermocouple or multiple thermocouples in the bed is essential in following this type of regeneration to control the oxygen supply and be sure the catalyst is not damaged.

Several techniques are used to attenuate the "hot spot" and lessen the chance of damaging the catalyst. A common method is to reduce the oxygen concentration to about 2–3% and recycle the gas. This has the effect of increasing the thickness of the hot spot and greatly reducing the peak temperature. The inert gases help carry away the heat of burning. To increase the overall rate of burning, the pressure of the recirculating gas is increased to about 2–10 atm.

It is well established that without secondary burning, carbon and graphite do not give 100% CO_2 but some CO, with the $CO:CO_2$ ratio increasing with temperature (*1*). A similar phenomenon occurs during the regeneration of some catalysts and 30–50% of the C may be removed as CO, thus reducing the heat liberated considerably. Generally, catalysts which have no oxidizing–reducing activity give this type of burning. In contrast, oxidizing–reducing catalysts such as Cr_2O_3 on alumina, Cr_2O_3 on silica–alumina, or Pt on alumina give little or no CO when coke is burned off them.

III. Burning in Single Granules

Weisz and Goodwin (*5, 6*) made a study and analysis of the coke burning in single catalyst granules under idealized conditions that illustrates some useful principles. They removed heat as it was generated, so that burning could be done at a constant temperature, and by using an excess of oxygen, the intrinsic rate of burning was determined. With a monolayer or less of coke, the intrinsic rate of burning was a first-order reaction:

$$d(C)/dt = -k(C) \tag{2}$$

where (C) is the concentration of coke, t is the time, and k is a constant that changes with oxygen concentration and temperature. Estimates are given indicating that each carbon atom in the deposit is as available as any other carbon atom at coke levels up to about 4% on a 100 $m^2\ gm^{-1}$ surface area alumina catalyst, or up to 6% on a 250 $m^2\ gm^{-1}$ silica–alumina catalyst. This tends to confirm calculations that such coke levels could indeed be present as monolayers.

When more than a monolayer of coke was clearly present, the burning rate was somewhat slower than Eq. (2) indicates and, as a monolayer was approached as a result of burning away the upper layers, the rate approached that of Eq. (2).

The energy of activation for the intrinsic burning rate was 37.6 kcal $mole^{-1}$. These relations held for 2 mm granules of several different catalysts and with deposits containing several different C:H ratios, leading one to expect rather general applicability to catalysts which do not have oxidizing–reducing activity.

The intrinsic rate constant found was:

$$k = 2 \times 10^8 \exp(-37{,}600/RT)\ \mathrm{sec^{-1}\ atm^{-1}} \tag{3}$$

where $k = (1/C)(dC/dt)$.

Reviewing the various studies on the effect of oxygen pressure, it has been tentatively concluded (*4*) that the reaction is first-order in oxygen, but the available data are not conclusive.

In the absence of better data the author would use Eq. (3) as a first approximation.

IV. Diffusion Limited Coke Burning

As Weisz and Goodwin (*5, 6*) increased the regeneration temperature above about 500°C, the intrinsic burning rate began to exceed the diffusion rate of O_2; the rate was more and more diffusion controlled as the temperature was increased further.

When the rate is diffusion controlled, some effects show up that were not present during intrinsic burning (*5, 6*). The burning proceeds from the outside surface of the granule toward the center so that a partially regenerated granule has a clean outside layer and an unregenerated core. The line of demarcation was the burning interface.

The $CO:CO_2$ ratio decreases as burning proceeds toward the center of the granule (*5, 6*). As oxygen diffuses through the pores to get to the burning interface, it has a chance to react with CO diffusing away from the burning. As the interface moves closer to the center of the granule, the probability of the exiting CO reacting with incoming oxygen increases, thus reducing the $CO:CO_2$ ratio. As would be expected, the extent of this secondary CO reaction is related to the diffusivity of the catalyst granules and the $CO:CO_2$ ratio decreases as the diffusivity decreases. Thus, the tightly pressed granules in Fig. 1, Chapter 1, are not desirable if minimum heat release is wanted during the coke burnoff. This is another reason for the optimum catalyst granule or bulk density.

V. Effect of Oxidizing–Reducing Activity

Oxidizing–reducing catalyst components, such as Cr_2O_3, MoO_x, Pd, Pt, etc., have a profound effect on the burning of catalyst deposits (*5, 6*). They increase the intrinsic burning rate and in one case it was two orders of magnitude faster than the uncatalyzed rate of Eq. (3). This permits lower temperature coke burnoffs at practical rates and is utilized with the Pt–alumina reforming catalysts (see Chapter 7) where regenerations can be conducted at 400°–450°C to minimize deactivation during regeneration. The fact that the rate of *carbon* burning could be accelerated by catalysts has been known for many years (*2, 3*).

These "combustion" catalysts change the nature of the burning reaction in that the $CO:CO_2$ ratio is reduced and is usually zero. The fact that the energy of activation of both types of burning is substantially the same may be fortuitous rather than an indication that the slow step is the same in both reactions (*5*, *6*).

VI. Burning Coke off Powdered Catalysts

Burning coke off powdered catalysts, as compared to granules, is relatively simple. The particles are usually so small that diffusion effects are absent. The burning is usually done in a fluidized bed, where any hot spots are immediately dissipated by the rapid movement of the catalyst, so that regeneration is done with air rather than with a gas containing 2–3% O_2.

The rate of burning and the $CO:CO_2$ ratio are subject to the presence or absence of oxidation–reduction activity just as with catalyst granules.

REFERENCES

1. Arthur, J. R., *Trans. Faraday Soc.* **47,** 164 (1959).
2. Day, J. E., and Robey, R. F., *Ind. Eng. Chem.* **28,** 564 (1936).
3. Day, J. E., Robey, R. F., and Dauben, H. J., *J. Am. Chem. Soc.* **57,** 2725 (1935).
4. Walker, P. L., Jr., Rusinko, F., Jr., and Austin, L. G., *Advan. Catalysis* **11,** 133 (1959).
5. Weisz, P. B., *J. Catalysis* **6,** 425 (1966).
6. Weisz, P. B., and Goodwin, R. B., *J. Catalysis* **2,** 397 (1963); **6,** 227 (1966).

CHAPTER 3

ISOMERIZATION

I. Isomerization of Paraffin Hydrocarbons

The main need for isomerizing paraffins is to convert *n*-paraffins into branched chain ones rather than the reverse. Isobutane is wanted to react with gaseous olefins (alkylation) to produce good antiknock motor fuel components. *n*-Pentane and *n*-hexane are converted directly into isomers which are good antiknock motor fuel components, much better than the *n*-paraffins.

Paraffin isomerization is an equilibrium process in which equilibrium is far from zero or 100% of any isomer under available operating conditions with currently known catalysts. Therefore, isomerization is combined with a separation process which segregates the poor antiknock hydrocarbons for recycle. With butane and pentane this process is distillation. With hexanes and higher paraffins distillation becomes expensive because the isomers boil too closely together for efficient separation. Molecular sieves which can separate the *n*-paraffins, i.e., the poorest antiknock components, are attractive alternatives.

The isomerization of *n*-paraffins is closely related to cracking. This is not a problem with butane and pentane; it begins to show up with *n*-hexane and is everpresent with *n*-heptane and higher paraffins. The cracking leads to rapid catalyst deactivation and excessive gas formation. For this reason, and because of the difficulty of separating the high antiknock components, isomerization is not ordinarily employed on the higher paraffins.

Because low temperatures favor the branched chain paraffins, the lowest practical operating temperatures are desired. This simplifies the separation problem because the desired isomer is present in a higher concentration. However, the available catalysts which operate at these low temperatures are strong acids (usually based on anhydrous aluminum chloride) and cause corrosion in the plant. The noncorrosive catalysts require higher operating temperatures and hydrogen pressure, so the equilibrium is less favorable. These two processes will be discussed separately, and since the second one requires hydrogen, it will be referred to as "hydroisomerization."

A. $AlCl_3$ Isomerization of Paraffin Hydrocarbons

The anhydrous $AlCl_3$ used as the catalyst is "promoted" or activated with anhydrous HCl. The processes differ primarily in the way in which the solid $AlCl_3$ is used in the system and the spent catalyst sludge is removed.

In one process, the feed stream is split and a part is passed through a bed of anhydrous $AlCl_3$ to dissolve the necessary amount. The two streams are combined and HCl added as the mixture passes through a bed of inert granules, e.g., quartz chips. An active HCl–$AlCl_3$–hydrocarbon liquid complex is formed which is supported in the bed, causing isomerization to occur. As the catalyst loses activity, it percolates toward the bottom of the bed where it is removed as a thick brown-black liquid. HCl is recovered and recycled. For butane isomerization, typical conditions are: 90°–100°C, 13–17 atm, LHSV 0.5 $V_l V_c^{-1}$ hr^{-1} (based on the volume of inert granules), HCl 5% of hydrocarbon. The conversion is 38% per pass and 400–1000 liters of isobutane produced per kilogram of $AlCl_3$ (*19*).

In a second version of the process, anhydrous $SbCl_3$ is used to dissolve the $AlCl_3$ so that a liquid catalyst is formed with the hydrocarbons. This is used in a stirred reactor (continuous throughput) to cause isomerization. For butane isomerization, typical conditions are 80°–90°C, 20 atm, LHSV 2.5 $V_l V_c^{-1}$ hr^{-1}, 5% HCl. Conversion is 45% per pass and isobutane production 400–1000 liters per kilogram $AlCl_3$ (*2*, *19*, *29*).

In a third version of the process, the HCl–$AlCl_3$–hydrocarbon liquid complex is used as the catalyst in a stirred reactor. For butane, conditions are 100°C, 25 atm, 4% HCl, and LHSV 1 $V_l V_c^{-1}$ hr^{-1} (*19*).

It will be recalled that $AlCl_3$ sublimes fairly readily. It is possible to sublime $AlCl_3$ onto an adsorptive support (bauxite or activated charcoal). This has been used as a catalyst with operating conditions similar to those above. Since regeneration is not feasible, the entire bed must be replaced periodically (*6*).

In all the above processes a small amount of hydrogen is produced, and this along with the HCl provides the pressure. The hydrogen helps control the reaction and reduces catalyst consumption.

Pentane and hexane isomerization technology is similar to butane isomerization. Here, the process can be complicated by a disproportionation reaction, e.g.:

$$2\ C_5H_{12} \rightarrow i\text{-}C_4H_{10} + i\text{-}C_6H_{14}$$

which consumes catalyst. A hydrogen partial pressure of about 4–7 atm is maintained to inhibit this reaction in pentane isomerization. In hexane processes, this pressure may reach 45–55 atm and the hydrogen consumption 7–15 liters per liter of pentane. With pentane, the small amount of hydrogen generated in the process may be enough.

In one instance, 0.5% benzene was included in the feed to control side reactions (*24*) but, in general, aromatic hydrocarbons tend to react with $AlCl_3$ and deactivate it for paraffin isomerization, thus increasing catalyst consumption. Therefore, higher concentrations are usually removed from the feed. Water and olefins also react with $AlCl_3$ and increase consumption.

The energy of activation of isomerization with $AlCl_3$ catalysts is about 10 kcal mole^{-1}.

No practical way exists to regenerate the catalyst sludges formed in these processes short of a complete chemical reworking, which is expensive. It is usually cheaper to replace the catalyst with fresh $AlCl_3$.

B. Hydroisomerization of Paraffin Hydrocarbons

Paraffin isomerization is one of the reactions that takes place in catalytic reforming, using the dual function Pt–alumina catalysts (see Chapter 7). Any of the commercial Pt–alumina reforming catalysts will isomerize paraffins but special modifications have been made to enhance activity for isomerization. Pt, Pd, or Ni seem to be the commonest hydrogenation components with silica–alumina (including the hydrogen + rare earth form of faujasite, Linde Type Y, molecular sieves) fluorided and chlorided alumina the commonest acidic components (*11*, *23*, *27*).

Operating temperatures are in the 100°–300°C range, pressure 5–40 atm, H_2:hydrocarbon mole ratios 1:4. Hydrogen consumption is 4–8 liters per liter of liquid feed. The catalyst is not corrosive and is not sensitive to aromatic hydrocarbons since they are hydrogenated by the catalyst to cyclohexanes which, in turn, are isomerized. Sulfur compounds poison the hydrogenation component of the catalyst so desulfurization of the feed is necessary. Wet feed ($H_2O > 100$ ppm) interferes with the active sites and competes with the hydrocarbon for them.

Conditions for the isomerization of pentane are 120°–175°C, 15–40 atm, and LHSV 1–4 $V_lV_c^{-1}$ hr^{-1}. For mixtures containing both pentane and hexane, conditions are 50°–200°C, 35–100 atm, and LHSV 1–3 $V_lV_c^{-1}$ hr^{-1} (*14*).

There is an isomerization process that seems to be in between isomerization and hydroisomerization. It resembles the $AlCl_3$ process in that an isomerizing surface layer is generated on a Pt–alumina catalyst by reacting an organic chloride with the Pt–alumina catalyst of the reforming type (*10*). The feed hydrocarbons are desulfurized and the H_2S plus any water, removed. The aromatics are hydrogenated to cycloparaffins. Isomerizing conditions are 95°–160°C, 17 atm, H_2:hydrocarbon mole ratio 2:1, and LHSV 1–2 $V_l V_c^{-1}$ hr^{-1}. The process period is long, several months. The HCl which is used in the process is recovered and recycled.

Apparently the activity of the catalyst depends on the pretreatment. For the Sinclair–Baker Pt–alumina catalyst (Engelhard Industries), the catalyst was first reduced in hydrogen at 480°C, then given an HCl treatment at 500°–550°C to add about 2% Cl^-, and then with thionyl chloride at 300°C to give a total Cl^- content of about 4%. Not all Pt–alumina catalysts responded equally to this type of activation (*18*).

The energy of activation for the hydroisomerization is around 20 kcal $mole^{-1}$.

The author is not acquainted with a regeneration procedure for these isomerization catalysts.

Attempts have been made to isomerize paraffin waxes to make lubricating oils or to reduce the freezing point of petroleum products. For example, higher *n*-paraffins have been studied with Pt–fluorided alumina and some degree of success attained, accompanied, however, by an undesirable amount of cracking (*26*). The author is not aware of a current commercial process based on this type of isomerization.

C. Suppliers of Proven Catalysts

Linde Division of Union Carbide Corp.

SK–120: a Pd on Linde Type Y molecular sieve for the hydroisomerization of pentane and hexane, as 3 mm tablets. The ABD is 0.75 and the surface area 550 m^2 gm^{-1}. The catalyst must be activated with hydrogen before use.

Universal Oil Products Co.

Type I–3: Pt on alumina for isomerizing pentane and hexane.

Type I–4: 0.375% Pt on spherical alumina for isomerizing butane.

II. Cycloparaffins

Like paraffins, cycloparaffins can be isomerized with anhydrous $AlCl_3$ + HCl or hydroisomerized with halided Pt–alumina catalysts (*11*). The conditions already given for paraffin hydrocarbons will work for the cycloparaffins. As with the paraffins, the reaction comes to equilibrium some distance from zero or 100% of any isomer, so a separation and recycle process is necessary for complete conversion.

Because the catalysts are the same as for paraffin isomerization, the poisons are the same.

Trimethylenenorborane isomerizes with $AlCl_3 + HCl$ or $AlBr_3 + HBr$, through a complex series of reactions at room temperature, to adamantane (*30*). In a similar way, six methyltrimethylenenorboranes isomerize to 1-methyladamantane (*31*). Perhydroacenaphthene gives 1-ethyl- and then 1,3-dimethyladamantane. Perhydrofluorene gives 1-ethyl-3-methyladamantane and then 1,3,5-trimethyladamantane. Perhydroanthracene gives 1,3-dimethyl-5-ethyladamantane and then 1,3,5,7-tetramethyladamantane (*32*).

III. Isomerization of Olefin Hydrocarbons

Olefin hydrocarbons can be isomerized in two ways: (*a*) move the double bond along the chain and (*b*) change the branching of the chain. In addition, cis–trans changes can take place in some olefins.

Almost any of the moderately acidic catalysts, e.g., the cracking catalysts (Chapter 4) or the acidic aluminas used with Pt as reforming catalysts (Chapter 7) will move the double bond along the chain (*11*). The reaction is not as simple as it seems. With a silica–alumina cracking catalyst and 1-butene, it has been shown that a certain amount of the olefin polymer forms on the catalyst, followed by 2-butene formation (*16*).

With branched chain olefins, both polymer formation and movement of the side chain can preclude a clean reaction. When the simple movement of a double bond was wanted in the following reaction, a special catalyst was developed to minimize side reactions.

$$CH_2{=}\overset{\displaystyle CH_3 \atop |}{C}{-}CH_2CH_2CH_3 \rightarrow CH_3\overset{\displaystyle CH_3 \atop |}{C}{=}CHCH_2CH_3$$

The catalyst was 4% NiO + 10% WO_3 on alumina (*25*). The reaction temperature is 150°C, the LHSV 10 $V_l V_c^{-1}$ hr^{-1}, and atmospheric pressure. On stream periods are about 8000 hours. The catalyst is used as part of a commercial isoprene synthesis. The 2-methylpentene-2 is cracked to isoprene and methane. The writer does not know a commercial supplier of the catalyst.

While so far as the author knows they are not being used commercially, the highly basic catalysts are more selective for double bond movement than the acidic ones. Sodium or lithium metal treated alumina is active at room temperature or slightly above (*20*). Potassium *t*-butylate in dimethyl sulfoxide is a homogeneous catalyst that is easier to prepare and is active in the 50°–100°C range (*33*).

There is still another way of moving the double bond along the chain, one that has considerable significance to the understanding of the intermediate formed between the olefin and the catalyst. If an olefin, e.g., 1-butene, is hydrogenated with Ni or Pd so that only part of the olefin is hydrogenated, it will be found that the double bond has moved on the unhydrogenated part (*8*). This is interpreted to mean that the olefin is adsorbed on the metal by opening the double bond.

$$1\text{-}C_4H_8 \rightarrow CH_3CH_2\underset{*}{C}H-\underset{*}{C}H_2$$

$\underset{*}{C}$ represents a carbon–metal bond. This reacts with chemisorbed $\underset{*}{H}$ to give $CH_3\underset{*}{C}H_2CH-CH_3$. This in turn loses a $\underset{*}{H}$ to give $CH_3\underset{*}{C}H-\underset{*}{C}HCH_3 + \underset{*}{H}$ which, on desorption, gives 2-butene. This picture not only provides an explanation, it also indicates that not all chemisorbed molecules are hydrogenated. This is further substantiated by the fact that cis–trans isomerization is faster than double bond migration.

A completely analogous reaction has been known for many years in the hydrogenation of the glyceride oils; unsaturated fat acid esters are produced during the hydrogenation which are isomers of the original and were not present in the starting material (*15*).

Not all hydrogenation metals are equally effective for double bond migration during hydrogenation. Iron, nickel, and palladium are good. Platinum and iridium are poor, while ruthenium, rhodium, and osmium are in between (*9*).

The reaction takes place in the liquid phase or in solution at room temperature. The finely divided catalyst is kept in suspension by stirring. For the noble metals, 5% metal on a support, e.g., activated carbon, is used. Nickel catalysts must be active at room temperature.

A. Suppliers of Proven Catalysts

Both Engelhard Industries and Matthey Bishop, Inc. supply the platinum group metals supported on activated carbon.

See Chapter 13 for nickel catalyst suppliers.

B. Chain Branching of Olefins

If one attempts to convert a straight chain olefin into a branched chain one, or vice versa, the branched olefin is partly polymerized, greatly reducing the yield (*11*). Interestingly, there is a fair amount of evidence indicating that it is an olefin intermediate that isomerizes in the paraffin hydroisomerization on Pt–alumina type catalysts discussed above. The Pt may dehydrogenate the paraffin to an olefin which moves to the acidic alumina where chain branching occurs, and the isomerized olefin moves back to the Pt where it is hydrogenated to the branched chain paraffin. Here hydrogenation takes place before significant polymerization can occur.

The cycloolefins also isomerize with acidic catalysts. For example, cyclohexene isomerizes to methylcyclopentenes at 250°C in the presence of the silica–alumina cracking catalyst (*13*). Again, however, the reaction is not a clean one and is accompanied by polymerization.

The author knows of no current commercial process for isomerizing olefins to branched isomers and cannot suggest a proven catalyst.

IV. Isomerization and Hydroisomerization of Alkylaromatic Hydrocarbons

It is well known that the silica–alumina cracking catalysts will equilibriate mixed xylenes. The catalyst loses activity fairly rapidly and has to be regenerated. By putting a little Pt on the silica–alumina and controlling its acidity, it is possible to keep the catalyst clean with hydrogen and to operate for long periods of time.

In isomerizing xylenes, usually to maximize *p*-xylene for preparing terephthalic acid, it is desirable to convert the ethylbenzene which is present into xylenes too. This is possible and, interestingly, a cycloparaffin intermediate seems to be involved. The temperature should be low enough so some of the ethylbenzene can be hydrogenated to provide the intermediate. Not all need be hydrogenated. For optimum utilization of ethylbenzene (or other aromatic hydrocarbons with more complicated side chains), a two-stage process can be used. Conditions for the first stage are 12 atm, 385°C, H_2:hydrocarbon mole ratio 10, and LHSV 2 $V_l V_c^{-1}$ hr^{-1} (*7, 28*). This is the partial hydrogenation stage. In the second stage, a dehydrogenation stage, the other conditions are the same but the temperature is 430°–500°C.

In a single stage operation where ethylbenzene conversion is satisfactory, suitable conditions are 12 atm, LHSV 1 $V_l V_c^{-1}$ hr^{-1}, H_2:hydrocarbon mole ratio 10, and temperature 430°–480°C.

In addition to that of moving methyl groups around on the benzene ring, the concept of chopping up a complex side chain and sticking the pieces on the ring is an interesting one. The observation that it goes through a cycloparaffin intermediate stage is certainly useful in deciding optimum operating conditions. The observation that isopropylbenzene can be converted into trimethylbenzenes and ethylmethylbenzenes using this technique is interesting but, so far as the author knows, is not used commercially.

A process using a catalyst that does not contain a noble metal has been developed for commercial operation (*1*). The operating temperature is 370°–450°C, the pressure 15–30 atm, the LHSV 0.5–3 $V_l V_c^{-1}$ hr^{-1}, and the hydrogen:hydrocarbon mole ratio 3:1–8:1. Neither the nature of the catalyst nor the name of the supplier is available to the author.

The xylenes are isomerized by the H-form as well as multivalent ion-exchanged forms of the Y type (Linde) molecular sieve and they are more active than the silica–alumina cracking catalysts (*21*). Evidence is available indicating that the isomerization with these multivalent ion-exchanged Y molecular sieve catalysts goes through a transalkylation stage in which trimethylbenzenes are intermediates (*22*).

Irrespective of the mechanism, the equilibrium concentration of *p*-xylene in the practical processes is about 24%.

A. Supplier of Proven Catalyst

Engelhard Industries supplies the Pt on silica–alumina of controlled acidity.

See Section I–G, Chapter 7, for Pt–alumina catalysts and Chapter 4 for suppliers of cracking catalysts.

V. Aromatic Carboxylic Acids

This process has been called both the Henkel and the Raecke process. Phthalic or isophthalic acid can be isomerized to terephthalic acid. Naphthalene dicarboxylic acids can be isomerized to the 2,6-dicarboxylic acid.

In one form, this may be called disproportionation instead of isomerization, for benzoic acid can be converted into terephthalic acid and benzene.

The potassium salt of the acid is dried, mixed with a small quantity of a cadmium salt (e.g., of the same acid), and the mixture heated to 400°–430°C in the presence of CO_2 at 3–8 atm pressure for the phthalates, or 10–15 atm for the benzoate (*4*).

Water and compounds containing an active hydrogen must be excluded. The system behaves almost as if metallic potassium were an intermediate so that things which react with it act as poisons.

Zinc may be used in place of cadmium as the catalyst but it is less effective, requiring higher CO_2 pressures. Sodium salts are far less effective than potassium salts.

The process has been operated on a limited scale commercially. It suffers from two major technical problems. The pasty mass is difficult to handle at reaction temperatures and the cost of recovering the potassium is high.

In a related process for making terephthalic acid from benzoic acid, a pressure of CO is used along with the CO_2. As a result, more than 0.5 mole of terephthalic acid is produced instead of the 0.5 mole in the Henkel process (*3*).

VI. Propylene Oxide to Allyl Alcohol

The catalytic conversion of olefin oxides into the corresponding carbonyl compounds by activated alumina at 200°–300°C is a well-known laboratory reaction. For example, ethylene oxide is converted into acetaldehyde, propylene oxide into propionaldehyde, etc. So far as the author is aware, the reaction is not used commercially since the olefin oxides are usually more valuable than the corresponding carbonyl compounds.

In contrast, the conversion of propylene oxide into allyl alcohol represents a selective reaction for making a product more valuable than the feed (*7*). Using a lithium phosphate catalyst at 230°–270°C, this reaction proceeds at atmospheric pressure with acceptable selectivity. At 250°C and an LHSV of 1 $V_l V_c^{-1}$ hr^{-1}, about 20% of the feed is converted per pass and the yield on converted material is better than 90% (*12, 17*).

The author is not aware of a commercial supplier of the catalyst. If the experimenter wants to make his own catalyst, the following directions are available (*12*). Dissolve 126 gm (3 gm moles) of lithium hydroxide monohydrate in 500 ml of water. Also dissolve 570 gm (1.5 moles) of sodium orthophosphate dodecahydrate in 750 ml of water and warm to 40°C. Over 1–2 minutes, add the phosphate solution to the lithium solution while agitating. Filter, wash the precipitate, and then disperse in 1.5 liters of water at 50°–95°C. Agitate for 30 minutes, filter, and wash. Repeat the leaching with hot water four times, then dry at 200°C. Granules of the catalyst are ready for use.

REFERENCES

1. Aalund, L. R., *Oil Gas J.* **66,** No. 29, 102 (1968).
2. Anonymous, *Hydrocarbon Process. Petrol. Refiner* **42,** No. 11, 206 (1963).
3. Anonymous, *Chem. Eng. News* **43,** No. 32, 30 (1965).
4. Anonymous, *Hydrocarbon Process.* **46,** No. 11, 229, 230 (1967).
5. Anonymous, *Hydrocarbon Process.* **47,** No. 9, 173 (1968).
6. Anonymous, *Hydrocarbon Process.* **47,** No. 9, 174 (1968).
7. Anonymous, *Oil Gas J.* **66,** No. 14, 80 (1968).
8. Bond, G. C., "Catalysis by Metals," p. 256. Academic Press, New York, 1962.
9. Bond, G. C., and Wells, P. B., *Advan. Catalysis* **15,** 91 (1964).
10. Burbidge, B. W., and Rolte, J. R. K., *Hydrocarbon Process.* **45,** No. 8, 168 (1966).
11. Condon, F. E., *Catalysis* **6,** 43 (1958).

12. Denton, W. I., U.S. Patent 2,986,585 (May 30, 1961).
13. Egloff, G., Morrell, J. C., Thomas, C. L., and Bloch, H. S., *J. Am. Chem. Soc.* **61,** 3571 (1939).
14. Erickson, R. A., and Asselin, G. F., *Chem. Eng. Progr.* **61,** No. 3, 53 (1965).
15. Feuge, R. O., *Catalysis* **3,** 432 (1955).
16. Finch, J. W., and Clark, A., *Am. Chem. Soc. Div. Petrol. Chem. Preprints* **13,** No. 3, 200 (1968).
17. Fowler, G. W., and Fitzpatrick, J. T., U.S. Patent 2,426,264 (August 26, 1947).
18. Giannetti, J. P., and Sebulsky, R. T., *Am. Chem. Soc. Div. Petrol. Chem. Preprints* **14,** No. 1, 97 (1969).
19. Gunness, R. C., *Advan. Chem. Ser.* **5,** 109 (1951).
20. Haag, W. O., and Pines, H., *J. Am. Chem. Soc.* **82,** 387 (1960).
21. Hansford, R. C., and Ward, J. W., *J. Catalysis* **13,** 316 (1969).
22. Lanewala, M. A., and Bolton, A. P., "The Isomerization of the Xylenes Using Zeolite Catalysts," Meeting of the Catalysis Society, Atlantic City, N.J., Feb. 21–22, 1969.
23. Lanewala, M. A., Pickert, P. E., and Bolton, A. P., *J. Catalysis* **9,** 91 (1967).
24. Mavity, J. M., Pines, H., Wackher, R. C., and Brooks, J. A., *Ind. Eng. Chem.* **40,** 2374 (1948).
25. Menapace, H. R., and Smith, R. R., U.S. Patent 3,268,617 (Aug. 23, 1966).
26. Orkin, B. A., *Ind. Eng. Chem. Prod. Res. Develop.* **8,** 154 (1969).
27. Pickert, P. E., Bolton, A. P., and Lanewala, M. A., *Chem. Eng.* **75,** No. 16, p. 133 (1968).
28. Pitts, P. M., Jr., Connor, J. E., Jr., and Leum, L. N., *Ind. Eng. Chem.* **47,** 770 (1955).
29. Ruedisulj, W. M. J., Evans, H. D., and Fountain, E. B., *Hydrocarbon Process. Petrol. Refiner* **42,** No. 7, 125 (1963).
30. Schleyer, P. von R., and Donaldson, M. M., *J. Am. Chem. Soc.* **82,** 4645 (1960).
31. Schleyer, P. von R., and Nicholas, R. D., *Tetrahedron Letters* p. 305 (1961).
32. Schneider, A., Warren, R. W., and Janoski, E. J., *J. Am. Chem. Soc.* **86,** 5365 (1964); also see Schneider, A., U.S. Patent 3,128,316 (April 7, 1964).
33. Schriesheim, A., Hofmann, J. E., and Rowe, C. A., *J. Am. Chem. Soc.* **83,** 3731 (1961).

CHAPTER 4

CATALYTIC CRACKING OF PETROLEUM

Petroleum fractions boiling from about 200° to over 500°C are used in catalytic cracking. In terms of the mass of material processed, catalytic cracking is the most important process, with about 900,000 tons per day processed in the U.S. as of this writing. In terms of solid catalyst requirement it is also the most important process in that some 400 tons of new catalyst are required each day.

In catalytic cracking, the fresh catalyst or freshly regenerated catalyst is quite active. However, in use, carbonaceous material, "coke," rapidly builds up on the catalyst, causing progressive deactivation and requiring frequent regeneration by burning off the coke. Since many useful catalytic techniques as well as catalysts have been and are employed, it is appropriate to review these in some detail even though there are many excellent summaries (*1*, *2*, *5–8*, *11*, *13*, *16–19*, *21*). It is first useful to define some terms. When the petroleum fractions come into contact with a cracking catalyst under cracking conditions, gas, gasoline, and coke are produced. Not all the petroleum fraction is converted to material outside the original boiling range. The part that is so converted is called "percent converted" and may be expressed as volume percent or weight percent. It is fully realized that there is material within the original boiling range that has been altered in the process but this is usually ignored. The distribution of the converted material among the several products is called "product distribution." A good product distribution usually means lots of gasoline in relation to gas and coke. Whether the products are olefinic, paraffinic, aromatic, excellent or poor antiknock in quality is "product quality." These various factors are controlled by operating conditions, the nature of the original petroleum fraction used, and the nature of the catalyst. The writer will not attempt to describe the interrelations here. Suffice it to say they are known, are complex, and are available on computer programs.

I. Catalytic Cracking Reactors

The early catalytic cracking reactors were fixed bed reactors with hot molten salt circulating in tubes within the bed to supply some of the heat of reaction. Because of the coke buildup during processing, the catalyst was used about 10 minutes before it was regenerated, and during this 10 minutes, the conversion was continuously declining. This meant that fractionating equipment to separate gasoline, etc., in the product was subject to ups and downs and so could not operate efficiently. As many as twelve separate reactors were operated in a precise program (*a*) to smooth out the irregularities and (*b*) to obtain the needed capacity. The feed material was vaporized so it could be appropriately distributed to the reactors. As the catalyst aged and lost activity, the productivity of the unit declined.

Next came the development of flowing the catalyst through the reactor so that constant conditions could be maintained there. This took two forms: (*a*) a moving bed of pellets and (*b*) a bed of fluidized powder.

In the moving bed, catalyst pellets of the same type used in the fixed bed were elevated to the top of the reactor and allowed to flow by gravity down through the reactor, countercurrent to the flow of the vaporized oil. The spent catalyst leaving the bottom of the reactor was moved to the regenerator. At first bucket elevators were used but later, air lifts were used to elevate the catalyst. Unusually hard catalyst pellets had to withstand the mechanical handling and even with the hardest catalysts, attrition did occur. Catalyst losses were about 0.1% by weight of the oil processed. Most of the oil going to the catalyst was vaporized before reaching the catalyst, but a minor part was injected as a liquid.

In the fluidized catalyst system, powdered catalysts were (and are) used that could be made to flow like a liquid as a result of a film of gas or vapor surrounding the particle. The entering liquid oil is injected into the hot catalyst from the regenerator and moved into the reactor. The catalyst is so active that a large part of the cracking takes place in the transfer tube into the reactor. As a result, in many units, the transfer tube has been lengthened and widened to produce a vertical tube ("riser") where most of the cracking occurs. At the same time, the size of the fluidized bed reactor has been reduced.

In the fixed bed reactor, the length of time oil vapor was in contact with catalyst was 10 minutes before regeneration. In the moving bed of pellets this was 5 minutes or less. In the fluidized reactors this is down to seconds or a minute or two. The weight of catalyst flowing through the reactor may be five, ten or more times the weight of oil flowing through. This means an enormous capability for moving the powdered catalyst. Some of the larger units move 100 tons or more of catalyst per minute.

With the large catalyst:oil ratios of the fluidized catalyst units, most of the heat required is supplied by hot catalyst coming from the regenerator. The oil is injected as a liquid and the hot catalyst furnishes the heat for vaporization and for cracking.

In the fixed bed the operating temperature was 450°C or a little higher. In the moving bed this moved up to around 500°C and with the fluidized catalyst, 525°–550°C. The LHSV in the fixed bed was around 1 $V_l V_c^{-1}$ hr^{-1} and in the moving bed, somewhat higher. Space velocity in the fluidized catalyst reactor begins to lose its meaning since V_c, the volume of catalyst, is ill defined. The WHSV has more meaning since the weight of catalyst in the reactor can be determined. It is more usual to define catalyst residence time (t, in minutes or hours) and catalyst:oil ratio as criteria of severity of the operation. WHSV can be calculated from these by the relation:

$$\text{WHSV} = \frac{1}{t \times C/O}$$

if t is expressed in hours. It is of the order of 10–20 $W_l W_c^{-1}$ hr^{-1}. Pressure is 2–4 atm.

As of this writing, there are no fixed bed catalytic cracking units operating in the U.S. There are many moving bed units operating. The major portion of the oil cracked is cracked in fluidized catalyst units.

It goes without saying that the petroleum fractions being cracked are complex mixtures of hydrocarbons. At the same time, it must be realized that not all the various hydrocarbons in the mixture have the same susceptibility to cracking. As a result, those hydrocarbons easiest to crack are converted first, leaving the more difficult ones. Truly aromatic hydrocarbons do not crack but pass through unchanged or condense on the catalyst to form coke. For this reason, the unconverted material can be improved by hydrogenating the aromatics to cycloparaffins which do crack (see Chapter 13).

Work with pure compounds indicates that olefins, if present, crack most readily, alkyl aromatics with more than two carbon atoms in a side chain dealkylate readily, the cycloparaffins are next, and the paraffins next.

At the same LHSV, or WHSV, the susceptibility to catalytic cracking seems to increase as the average boiling point of the petroleum fraction increases. This is more apparent than real, because the average molecular weight of the fraction with the lower average boiling point is lower, so a given volume or weight contains more molecules. Therefore, in unit time the catalyst may be cracking just as many molecules of the lower boiling fraction as the higher boiling one. With the higher boiling fraction, this number of molecules is a large fraction of the total number, so the volume or weight percent conversion is large. With the lower boiling fraction, this number of molecules is a smaller fraction of the total number, so the weight or volume percent conversion is small. In this instance, space velocity as a useful device in catalysis can lead to an ambiguous conclusion.

If small hydrocarbon molecules are mixed with large ones, the large ones may be preferentially adsorbed and crack while preventing the smaller ones from cracking, and again the small molecule appears to be more refractory.

II. Catalyst Regeneration

In all the cracking catalyst regenerations, steam is used to purge hydrocarbons from the catalyst so they can be recovered rather than burned during regeneration. This requires the catalyst to be reasonably stable to high temperature steam.

The rate at which coke is burned off the cracking catalysts is considerably slower than the cracking rate. One might say that from the beginning, catalytic cracking has been a catalyst regeneration process in which the development of regeneration techniques has dominated. The developments have included ways in which the heat evolved during regeneration could be used in the process.

In the fixed bed units, the time spent purging and regenerating was about 20 minutes, i.e., twice as long as the time spent processing. Molten salt was circulated in tubes inside the catalyst beds and this was heated during regeneration and served to carry the heat from regenerating reactors to processing reactors. Salt temperatures were limited to 485–500°C, although doubtless some of the catalyst pellets experienced higher temperatures. As a result of the regeneration temperatures, the catalyst particles shrank and the volume of the catalyst bed decreased. Before replacing the catalyst, reactors would be opened and fresh catalyst added to reestablish the original volume.

In the moving bed units, the catalyst moves hot from the regenerator to the reactor and so carries regeneration heat. The regenerator contains cooling coils in which steam is generated. The temperature of regeneration is 550°–575°C. The catalyst spends about 2 minutes in the regenerator for each minute spent in the reactor. Even with the cooling coils, sometimes hot spots develop that fuse the catalyst into large lumps; then the unit has to be shut down and the lumps removed.

In the fluidized catalyst regenerators, the temperature is 600°–650°C with a few units operating as high as 700°C (*12*). This increases the rate of burning. Operating pressure has also been increased so the partial pressure of oxygen in the regenerator is increased. This too has helped speed up burning, but the regenerator must be larger than the reactor because coke burnoff is still slower than formation. Due to the motion of the catalyst in the regenerator, there are no localized hot spots to fuse the catalyst.

Regeneration in all types of units has certain common features. Some of the carbon is removed as CO (see Chapter 2) and the proportion varies some with the catalyst and other factors. However, if the exit gas also contains a significant concentration of oxygen, the CO and oxygen can get together as "after burning." Uncontrolled, this can damage equipment, so a variety of techniques have been developed to detect it and to control it, usually by injecting water to extinguish the fire. Otherwise, the CO is taken to a furnace where it is burned to generate steam.

In none of the catalyst regeneration systems is the coke removed completely. As a result, the "regenerated" catalyst still contains a few tenths of a percent of coke. With improved regeneration conditions, the residual coke is in the 0.2–0.4% range based on the catalyst. With the fluidized catalyst it is possible to determine the coke content of the catalyst by visual comparison with a set of standards.

Sulfur compounds are more readily adsorbed by the catalyst than hydrocarbons. In any event, the catalyst coke contains a higher percentage of S than the feed. This results in some desulfurization in catalytic cracking. The sulfur in the coke is burned to SO_2 which is discharged to the atmosphere.

III. Cracking Catalysts

With minor exceptions, all commercial cracking catalysts are based on silica–alumina combinations of one type or another. Broadly, they can be divided into three classes: (1) the acid-treated natural aluminosilicates; (2) the amorphous synthetic silica–alumina combinations. (3) the crystalline, synthetic silica–alumina combinations. All of them are high temperature acids and their catalytic activity is attributed to this acidity. However, as of this writing the nature of the acidity has not been resolved; that is, whether it is Bronsted acidity, Lewis acidity, or both.

Even though Class 3 catalysts above account for 90% of the cracking catalysts currently being used, it is appropriate to consider the catalysts in a somewhat historical way.

The earliest cracking catalysts of this type were the acid-treated montmorillonite clays. These clays are hydrous aluminosilicates containing some base-exchangeable (zeolite type) ions. During the acid treatment, these zeolite ions as well as about one-half of the aluminum in the aluminosilicate structure are removed (*20*). The catalysts have been widely used but had two weaknesses. There was a certain amount of iron in the crystal lattice which, with high sulfur petroleums, became active. As a result, it would oxidize during regeneration and during the process period would catalyze the formation of coke and hydrogen. In addition, they were sensitive to high regeneration temperatures. Later, these weaknesses were more or less overcome in several ways. Other aluminosilicates were acid treated, e.g., halloysite and kaolinite. Also, clays were treated to extract more than one-half the alumina and some iron with it, after which some alumina was added back, i.e., a type of semisynthetic catalyst was made. Other methods were also used for making semisynthetic types.

The amorphous silica–alumina combinations were based on combining silica and alumina hydrogels so that the catalysts contained 10–15% Al_2O_3 or 20–30% Al_2O_3, "low" or "high" alumina catalysts, respectively. It was necessary to remove essentially all the alkali metal ions to obtain temperature and steam stability. These catalysts were iron-free, so they could be used with high sulfur oils. They were also more thermally stable so they could withstand high regeneration temperatures. They did have some deficiencies compared to the acid-treated montmorillonites. They gave a poorer product distribution but better product quality. Also, they gave lower conversions per pass before "overcracking" occurred. This latter term can be explained as follows. As one attempts to obtain higher conversions per pass by making conditions more severe, a region is reached in which gasoline is cracked to gas and coke; further increases in severity produce incremental conversion but no incremental gasoline. With the acid-treated montmorillonites, it was possible to go to higher conversions without overcracking than with the amorphous silica–aluminas.

Synthetic silica–magnesia catalysts have the good properties of the acid-treated montmorillonites and attempts have been made to utilize them or silica–alumina–magnesia. As of now, however, they have not gained wide acceptance.

The natural mineral faujasite and the synthetic Linde X and Y molecular sieves have a common characteristic crystalline structure. When the base-exchangeable ions are replaced in part by rare earth ions and in part by ammonium ions, an extremely active cracking catalyst is formed on heating to dehydrate and deammoniate the product (*9*, *14*, *15*).

The minute crystals of the processed molecular sieve are suspended in a matrix of another material and make up 8–10% or more by weight of the total catalyst. Each catalyst supplier has his own method of processing and matrix used. One matrix is the amorphous silica–alumina described above and this may be either the low or the high alumina type.

The molecular sieves are formed by a hydrothermal treatment of a highly alkaline silica–alumina composition. It is possible to treat a natural aluminosilicate clay under hydrothermal conditions, forming some molecular sieve domains in the clay structure. These may or may not be equivalent to a processed Linde X or Y type sieve in a matrix.

Broadly, these catalysts have four characteristics that differentiate them from the amorphous silica–alumina catalysts. (*A*) They are more active. (*B*) The gasolines produced contain a larger percentage of paraffinic and aromatic hydrocarbons than the gasolines from amorphous silica–aluminas. (*C*) It is possible to go to larger conversions per pass without overcracking. (*D*) They give a better product distribution (*4*).

IV. Catalyst Evaluation

It is necessary to have laboratory methods for evaluating catalysts on a small scale. These follow the activity of the catalyst in a plant to confirm whether the activity is holding, increasing, or decreasing. They are also useful for the preliminary evaluation of catalysts offered by suppliers. Only when a catalyst has passed such an evaluation, including heat and steam stability tests, will it be considered for plant use. These methods all use small laboratory cracking units operated under carefully controlled conditions so activity, product quality, and product distribution can be determined (*3*, *10*). The catalyst suppliers have their own test methods and will often make them available to users.

V. Poisons

Basic nitrogen compounds in the oil act as temporary poisons for the cracking catalysts. Apparently they react with the acid centers on the catalyst and inhibit the cracking reaction. They also add to the coke production. This does not mean that oils contain basic nitrogen cannot be cracked. Rather, it means that the apparent catalyst activity is lower than it would be in the absence of nitrogen compounds. 1000 ppm of nitrogen is considered high and 3500, very high.

Metallic compounds present in the oil deposit and accumulate on the catalyst. The most common ones are iron, nickel, and vanadium. They exert their own catalytic action in two ways. During processing, they catalyze the formation of coke and hydrogen without contributing to gasoline formation. On regeneration, they catalyze combustion to CO_2 rather than to CO, and so liberate unnecessary quantities of heat in the regeneration. Methods that remove these contaminants from the catalyst are available but have not gained wide acceptance. Fe is the least damaging of the three, V is about four times Fe in its effect, and Ni is about fourteen times as bad as Fe. If the metallic contaminants concentration in the catalyst are expressed as ppm, then the total damaging effect can be expressed:

$$\mathrm{Fe} + 4\,\mathrm{V} + 14\,\mathrm{Ni}$$

Cu, if present, is treated like Fe. When the sum of these is 1000 or more, the catalyst is badly contaminated.

Some users find catalysts containing molecular sieves less susceptible to metallic contamination than the amorphous silica–aluminas. In some cases, contamination as high as 2000 to 3000 ppm on the above scale may be tolerated.

When it is necessary to process oil with a very high metal content, an inexpensive catalyst such as a semisynthetic may be used so that the contaminated catalyst can be discarded and replaced faster than would otherwise be the case.

VI. Suppliers of Proven Catalysts

There is a large number of cracking catalysts and the reader can easily become confused. In an attempt to aid the reader, the author has grouped the catalysts into the three classes given above. He has selected the class for each catalyst based on the information available to him. The supplier may or may not agree with the classification, especially since it is somewhat oversimplified.

A. Class 1. Acid-Treated Natural Aluminosilicates and Semisynthetics

American Cyanamid Company

Aerocat 2000: a semisynthetic fluid catalyst containing 35% Al_2O_3. The ABD is 0.60, the surface area 300 m^2gm^{-1}, and the pore volume 0.50 cm^3gm^{-1}.

Davison Chemical Division of W. R. Grace & Co.

Grade SS: a semisynthetic fluid catalyst containing 32% Al_2O_3. It is supplied in two different pore volume grades: 0.58 cm^3gm^{-1} (ABD 0.51) and 0.70 cm^3gm^{-1} (ABD 0.47). The surface area of the latter is 280 m^2gm^{-1}.

Filtrol Corporation

Grade 58: 17.5% Al_2O_3, fluid. The ABD is 0.65, the surface area 280–300 m^2gm^{-1}, and the pore volume 0.36 cm^3gm^{-1}.

Grade 62: 17.5% Al_2O_3 as 3/16 × 3/16 inch pellets. The ABD is 0.8, the surface area 280–300 m^2gm^{-1}, and the pore volume 0.36 cm^3gm^{-1}.

Grade 63: 38% Al_2O_3 as 3/16 × 3/16 inch pellets. The ABD is 0.8, the surface area 125–135 m^2gm^{-1}, and the pore volume 0.27 cm^3gm^{-1}.

Grade 80: 38% Al_2O_3, fluid. The ABD is 0.73, the surface area 125–135 m^2gm^{-1}, and the pore volume 0.27 cm^3gm^{-1}.

Grade 100: 51% Al_2O_3, microspheres. The ABD is 0.80, the surface area 105 m^2gm^{-1}, and the pore volume 0.37 cm^3gm^{-1}.

Grade 110: pellets.

Grade 110: spherical pellets.

Houdry Process and Chemical Company

Kao-Pellets: approximately 3/16 × 3/16 inch. The ABD is 0.77 and the surface area 90–100 m:gm^{-1}.

Kao-Spheres: 45% Al_2O_3, ca. 0.17 inch in diameter. The ABD is 0.77 and the surface area 90–100 m^2gm^{-1}.

Nalco Chemical Company

Nalcat 783: a semisynthetic fluid catalyst, 33% Al_2O_3. The ABD is 0.50, the surface area 280 m^2gm^{-1}, and the pore volume 0.65–0.70 cm^3gm^{-1}.

B. Class 2. Amorphous Synthetic Silica–Alumina, Including Silica–Magnesia

American Cyanamid Company

Aerocat: 13% Al_2O_3, fluid. The ABD is 0.49 and the pore volume 0.75 cm^3gm^{-1}.

Aerocat Triple A: 25% Al_2O_3, fluid. The ABD is 0.43 and the pore volume 0.89 cm^3gm^{-1}.

Aerocat 3C–12: 3% MgO in low alumina, fluid.

Aerocat 3C–20: 3% MgO in high alumina, fluid.

Davison Chemical Division of W. R. Grace & Company

Low Alumina: 13% Al_2O_3, fluid. The ABD is 0.43 and the pore volume 0.77 cm^3gm^{-1}.

High Alumina: 28% Al_2O_3, fluid, in three different pore volumes: 0.70 cm^3gm^{-1} (ABD 0.46), 0.78 cm^3gm^{-1} (ABD 0.43), 0.88 cm^3gm^{-1} (ABD 0.39).

SM–30: 27.5% MgO and 3% F, fluid. The ABD is 0.49 and the pore volume 0.72 cm^3gm^{-1}.

Houdry Process and Chemical Company

S–46: 13% Al_2O_3, tablets. The ABD is 0.62, the surface area 280–315 m^2gm^{-1}, and pore volume 0.61 cm^3gm^{-1}.

Mobil Chemical Company

Durabead 1: 10% Al_2O_3, spheres (beads).

Nalco Chemical Company

Nalcat Low Alumina: 13% Al_2O_3, fluid. The ABD is 0.40, the surface area 520 m^2gm^{-1}, and the pore volume 0.80–0.85 cm^3gm^{-1}.

Nalcat High Alumina: 25% Al_2O_3, fluid. The ABD is 0.40–0.44, the surface area 440 m^2gm^{-1}, and the pore volume 0.8–0.9 cm^3gm^{-1}.

Universal Oil Products Company

Type FC–2: 13% Al_2O_3, fluid.

Type FC–3: 25% Al_2O_3, fluid.

C. Class 3. Crystalline Synthetic Silica–Alumina Combinations

American Cyanamid Company

Aerocat S–4: contains rare earth exchanged "Y" type molecular sieve in a semisynthetic matrix of 33% Al_2O_3 content, fluid. The ABD is 0.53, the surface area 330 m^2gm^{-1}, and the pore volume 0.57 cm^3gm^{-1}.

Aerocat TS–150: contains rare earth exchanged "Y" type molecular sieve in a matrix of synthetic silica–alumina (15% Al_2O_3), fluid. The ABD is 0.49, the surface are 600 m^2gm^{-1}, and the pore volume 0.65 cm^3gm^{-1}.

Aerocat TS–170 and TS–260: contain rare earth exchanged "Y" type molecular sieve in a semisynthetic matrix with approximately 33% Al_2O_3 content, fluid. The ABD is 0.55 and the pore volume 0.58 cm^3gm^{-1}.

Davison Chemical Division of W. R. Grace & Co.

XZ–15: 13% Al_2O_3, fluid. The ABD is 0.40, the surface area 500 m^2gm^{-1}, and the pore volume 0.88 cm^3gm^{-1}.

XZ–25: 36% Al_2O_3, fluid. The ABD is 0.5, the surface area 340 m^2gm^{-1} and the pore volume 0.60 cm^3gm^{-1}.

XZ–36: 36% Al_2O_3, fluid. The ABD is 0.55 and the pore volume 0.55 cm^3gm^{-1}.

XZ–40: fluid.

Filtrol Corporation

Grade 800: microspheres, 48% Al_2O_3. The ABD is 0.69, the surface area 210 m^2gm^{-1}, and the pore volume 0.39 cm^3gm^{-1}.

Grade 810: pellets.

Houdry Process and Chemical Company
HZ–1: pellets.

Mobil Chemical Company
Durabead 6B and Durabead 8: as spheres (beads); D–5 and D–7: fluid.

Nalco Chemical Company
KSF Series: "X" type molecular sieve in a matrix, fluid.
KSG Series: "Y" type molecular sieve in a matrix, fluid.
ND–2: low surface area, "Y" type molecular sieve in a matrix, fluid.

REFERENCES

1. Anonymous, *Oil Gas J.* **66,** No. 14, 103 (1968).
2. Ardern, D. B., and Dart, J. C., *Advan. Chem. Ser.* **5,** 13 (1951).
3. Ciapetta, F. G., and Henderson, D. S., *Oil Gas J.* **65,** No. 42, 88 (1967).
4. Eastwood, S. C., Drew, R. D., and Hartzell, F. D., *Oil Gas J.* **62,** No. 44, 158 (1962).
5. Greensfelder, B. S., *Advan. Chem. Ser.* **5,** 3 (1951).
6. Haensel, V., *Advan. Catalysis* **3,** 179 (1951).
7. Hansford, R. C., *Advan. Catalysis* **4,** 1 (1952).
8. McKetta, J. J., Jr., *Advan. Petrol. Chem. Refining* **2,** 580 (1959); **5,** 211 (1962).
9. Miale, J. N., Chen, N. Y., and Weisz, P. B., *J. Catalysis* **6,** 278 (1966).
10. Mills, G. A., Ashwill, R. E., and Gresham, T. L., *Hydrocarbon Process.* **46,** No. 9, 121 (1967).
11. Murphree, E. V., *Advan. Chem. Ser.* **5,** 30 (1951).
12. National Petroleum Refiners Association Forum, *Oil Gas J.* **67,** No. 6, 86 (1969).
13. Oblad, A. G., Milliken, T. H., Jr., and Mills, G. A., *Advan. Catalysis* **3,** 179 (1951).
14. Plank, C. J., and Rosinski, E. J., *Chem. Eng. Progr.* **63,** 26 (1967).
15. Plank, C. J., Rosinski, E. J., and Hawthorne, W. P., *Ind. Eng. Chem. Prod. Res. Develop.* **3,** 165 (1964).
16. Prater, C. D., and Lago, R. M. *Advan. Catalysis* **8,** 293 (1956).
17. Ries, H. E., Jr., *Advan. Catalysis* **4,** 87 (1952).
18. Ryland, L. B., Tamele, M. W., and Wilson, J. N., *Catalysis* **7,** 1 (1960).
19. Shankland, R. V., *Advan. Catalysis* **6,** 271 (1954).
20. Thomas, C. L., Hickey, J., and Stecker, G., *Ind. Eng. Chem.* **42,** 866 (1950).
21. Voge, H. H., *Catalysis* **6,** 407 (1958).

CHAPTER 5

DEHYDRATION

Dehydration, especially of alcohols, is one of the oldest catalytic reactions. A great many organic compounds alone or in mixture with other substances have been dehydrated with many different catalysts. It is not the purpose of this chapter to review these since this has been done thoroughly (*9*). For the reader with the problem of removing the elements of water from an organic system, a study of the review will be well worthwhile.

The purpose of this chapter is to provide enough background to support the proven catalysts at the end of the chapter.

I. Dehydration of Alcohols to Olefinic Hydrocarbons

The catalytic dehydration of ethyl alcohol to ethylene is a convenient source where only moderate quantities are needed. A great many catalysts have been used ranging from clays (kaolin) to γ-alumina and silica–alumina. Both γ-alumina and silica–alumina are good reliable catalysts. Typical operating conditions are 250°–350°C, atmospheric pressure or slightly higher, and an LHSV of 1–4 $V_l V_c^{-1}$ hr^{-1}.

The reaction is general except for those alcohols which cannot structurally give an olefin (many of these give ethers). As one goes to butyl alcohol and higher, more care must be used in selecting the catalyst. From *n*-butyl alcohol one expects to get 1-butene. However, if the catalyst has an acidic nature, some 2-butene will be formed. One commercial alumina (Harshaw Al–0104 T) has evidence in the literature (*5*) that it is nonacidic. As noted in the proven catalysts below, other aluminas contain alkali to overcome any acidity. It is worth noting that a 2-alkanol which one would expect to give mostly a 2-alkene actually gave a 1-alkene with a thoria catalyst (*3*).

While tungsten oxide and sulfide are components of hydrogenation catalysts and might be expected to be dehydrogenation catalysts for alcohols, actually blue W_2O_5 is a very active dehydration catalyst, as shown long ago by Sabatier (*8*). Commercial catalysts containing WO_3 (reduced under operating conditions) are available specifically for dehydration.

In many ways, the dehydration of an alcohol to an olefin acts as if the ether were an intermediate. By lowering the temperature some 50°C, significant yields of ether can be obtained with the primary alcohols. Ethers of secondary alcohols are more difficult to obtain and they are not formed from the tertiary alcohols.

While nickel in the presence of hydrogen is thought of as a hydrogenation or dehydrogenation catalyst, it can act as a dehydration catalyst to convert some alcohols to ethers. The nickel must contain some nickel oxide (about 2%) (*6*) and in precipitating the nickel hydroxide (or carbonate) ammonia (or ammonium carbonate) must be used instead of alkali. The catalyst can be supported on kieselguhr (7), but this is not essential. The nickel hydroxide or carbonate must be decomposed and reduced by hydrogen before use. At temperatures in the range of 125°–190°C and hydrogen pressure of 4 atm in a pulse microreactor, *n*-butyl, isobutyl, neopentyl, and 2- and 3-methylbutyl alcohols gave the corresponding ethers in yields ranging from 60 to 88% at conversions per pass up to about 50%. The selectivity for ether formation decreased as conversion per pass increased in the range of 10–50% (7).

sec-Butyl alcohol gave a 30% yield of the ether but *ter*-butyl alcohol gave only hydrocarbons.

Although the author does not know of a commercial supplier for this specific catalyst, the reaction is included here because it provides a route to some ethers (e.g., dineopentyl ether) not provided by the other dehydration catalysts.

A. Higher Alcohols to Olefins

Not all dehydration catalysts are heterogeneous. A laboratory method for converting diacetone alcohol into mesityl oxide uses a catalytic quantity of iodine (0.01%) dissolved in the alcohol. The solution is heated in a distillation flask so that the water and mesityl oxide distill as formed (*4*).

A similar technique can be used with powdered heterogeneous catalysts and alcohols that dehydrate readily. For example, 5–10% powdered silica–alumina or acid-treated montmorillonite (previously heated to 450°C) suspended in cyclohexyl alcohol and refluxed (140°–160°C) causes dehydration, so cyclohexene and water can be distilled as formed.

In dehydrating an alcohol to an olefin, unless the user has good reasons for some other catalyst, the author strongly recommends starting with a nonacid γ-alumina.

II. Dehydration of Glycols to Dienes

1,3-Butanediol dehydrates to 1,3-butadiene with the catalysts mentioned above and under the same conditions. 1,4-Butanediol also dehydrated to butadiene but via tetrahydrofuran as an intermediate. 2,3-Butanediol, on the other hand, goes to methyl ethyl ketone. Glycol goes to acetaldehyde. In contrast, 2,3-dimethylbutane-2,3-diol goes to dimethylbutadiene.

Butenols that might be considered intermediates also dehydrate to butadienes.

A somewhat related reaction is the catalytic decomposition of 4,4-dimethyldioxane to isoprene, water, and formaldehyde (*1*).

A. Poisons

The dehydration catalysts are not susceptible to poisons likely to occur in the feed material. They gradually accumulate coke deposits over a period of hundreds of hours and have to be regenerated.

B. Regeneration

The dehydration catalysts regenerate readily with air as long as the heat of regeneration can be dissipated so that the maximum temperature does not exceed 600°–650°C (*2*).

III. Suppliers of Proven Catalysts

Aluminum Company of America

Grade F–1: 92% Al_2O_3, 6.8% volatile, 0.8% Na_2O, 0.12% Fe_2O_3, as granules. The ABD is 0.8, the surface area 210 m^2gm^{-1}, and the pore volume 0.25 cm^3gm^{-1}.

Grade F–20: a gamma alumina, 92% Al_2O_3, 6.8% volatile, 0.8% Na_2O, 0.12% Fe_2O_3, as powder. The ABD is 0.95 and the surface area 210 m^2gm^{-1}.

Grade H–51: 85% Al_2O_3, 6.0% volatile, 5.8% SiO_2, 2.0% Na_2O, 0.1% Fe_2O_3, as microspheres. The ABD is 0.8, the surface area 350 m^2gm^{-1}, and the pore volume 0.5 cm^3gm^{-1}.

Grade H–151: 85% Al_2O_3, 6.2% volatile, 6.3% SiO_2, 2.0% Na_2O, 0.15% Fe_2O_3, as ⅛ and ¼ inch spheres. The ABD is 0.8, the surface area 350 m^2gm^{-1}, and the pore volume 0.3 cm^3gm^{-1}.

Davison Chemical Division of W. R. Grace & Co.

Special silica–alumina: 87% SiO_2, 13% Al_2O_3 for dehydration. Grade 970: the ABD is 0.67, the surface area 95 m^2gm^{-1}, and the pore volume 0.34 cm^3gm^{-1}, as granules.

Filtrol Corporation

Grades 13, 24, 25 and 49: a series of acid activated clays (aluminosilicates). These are "acidic" catalysts.

Grade 86: 7% volatile, 0.07% Fe_2O_3 (volatile-free basis), 0.02% Na_2O and 3.5% SO_3, as ⅛ and 3/16 inch extrudates. The ABD is 0.7 and the surface area 300–325 m^2gm^{-1}.

Grade 90: 20–26% volatile, 0.05% Fe_2O_3 (volatile-free basis), 0.02% Na_2O, 3.5% SO_3, as microspheres. The surface area is 300 m^2gm^{-1}.

Harshaw Chemical Company

Al–0104 T: activated alumina, 99% Al_2O_3, as ⅛, 5/32, 3/16, and ¼ inch tablets. The ABD is 0.7–1.0, the surface area 80–100 m^2gm^{-1}, and the pore volume 0.28 to 0.33 cm^3gm^{-1}.

Al–1401 P: active alumina, 97% Al_2O_3, as microspheres.

Al–1404 T: active alumina, 96% Al_2O_3, as ⅛, 5/32, 3/16, ¼ and ⅜ inch tablets. The ABD is 0.75–0.90, surface area 180–200 m^2gm^{-1}, and the pore volume 0.42–0.51 cm^3gm^{-1}.

Al–1602 T: silicated alumina, 91% Al_2O_3 and 6% SiO_2, as ⅛ inch tablets. The ABD is 0.85, the surface area 210–240 m^2gm^{-1}, and the pore volume 0.48 cm^3gm^{-1}.

Al–1605 P: silicated alumina, 92% Al_2O_3 and 6% SiO_2, as powder. The ABD is 1 and the surface area 210–240 m^2gm^{-1}.

Al–1706 E: active alumina, 97% Al_2O_3, as ⅛ inch extrudates. The ABD is 0.5, the surface area 218 m^2gm^{-1}, and the pore volume 0.77 cm^3gm^{-1}.

Al–2810 P: active alumina, 98% Al_2O_3, $<$0.05% Na_2O, 20% volatile, as powder. The ABD is 0.45, the surface area 340 m^2gm^{-1} and the pore volume 0.76 cm^3gm^{-1}.

Al–3404 P: active alumina, 99+% Al_2O_3, $<$0.008% Na_2O, 10% volatile, as powder. The ABD is 1.0, the surface area 175–200 m^2gm^{-1}, and the pore volume 0.40 cm^3gm^{-1}.

W–0101 T: ca. 90% WO_3 and 10% Al_2O_3, as ⅛ inch tablets. The ABD is 1, the surface area 75 m^2gm^{-1}, and the pore volume 0.37 cm^3gm^{-1}.

W–0801 T: ca. 90% WO_3 and 10% Al_2O_3, as tablets. The ABD is 1, the surface area 145 m^2gm^{-1}, and the pore volume 0.36 cm^3gm^{-1}.

Houdry Process and Chemical Company

Grade 100: 98.5% Al_2O_3, 0.4 to 0.5% Na_2O, as 3⁄32, 1⁄8, 5⁄32 and 3⁄16 inch extrudates. ABD is 0.78–0.82 and surface area 75–85 m^2gm^{-1}.

Grade 100 S, 98.5% Al_2O_3, 0.1–0.2% Na_2O, as 3⁄32, 1⁄8, 5⁄32 and 3⁄16 inch extrudates. ABD is 0.78–0.82 and surface area 75–85 m^2gm^{-1}.

Grade 200, 90.5% Al_2O_3, 7–8% bound H_2O, 0.4–0.5% Na_2O, as 3⁄32, 1⁄8, 5⁄32 and 3⁄16 inch extrudates. ABD is 0.82–0.86 and surface area 180–210 m^2gm^{-1}.

Grade 200S, 98% Al_2O_3, 0.1–0.2% Na_2O, as 3⁄32, 1⁄8, 5⁄32 and 3⁄16 inch extrudates. ABD is 0.77–0.81 and surface area 150–180 m^2gm^{-1}.

See Chapter 4 for other silica–alumina catalysts.

REFERENCES

1. Anonymous, *Hydrocarbon Process.* **46,** No. 11, 193 (1967).
2. Levy, R. M., Bauer, D. J., and Roth, J. F., *Ind. Eng. Chem. Prod. Res. Develop.* **7,** 217 (1968).
3. Lundeen, A. J., and van Hoozer, R., *J. Am. Chem. Soc.* **85,** 2180 (1963).
4. *Org. Syn.* **1,** 347 (1941).
5. Pines, H., and Haag, W. O., *J. Am. Chem. Soc.* **82,** 2471 (1960).
6. Pines, H., and Kobylinski, T. P., "Hydrogenolysis of Alcohols. II. Reduced Nickel Oxide—Its Intrinsic Acidity and Catalytic Activity in The Conversion of Alcohols to Ethers." Meeting of the Catalysis Society, Atlantic City, N.J., Feb. 20–22, 1969.
7. Pines, H., and Steingaszner, P., *J. Catalysis* **10,** 60 (1968).
8. Sabatier, P., and Mailhe, A., *Ann. Chim. Phys.* **20,** 328 (1910); also see Emmett, P. H., Sabatier, P., and Reid, E. E., "Catalysis Then and Now," p. 257. Franklin Publ. Co., Englewood, New Jersey, 1965.
9. Winfield, M. E., *Catalysis* **7,** 93 (1960).

CHAPTER 6

DEHYDROGENATION

Both of the famous pioneers in catalysts, V. N. Ipatieff and Paul Sabatier, described useful dehydrogenation catalysts and reactions. Some of them form the foundation for the proven rugged catalysts now available. These early investigators hardly dreamed of the wide utility of their catalysts.

Although dehydrogenations are strongly endothermic, the heat of reaction is usually supplied by preheating the reactant and charging it to a fixed bed adiabatic reactor. As a result, the temperature in the catalyst bed drops in the direction of flow. For this reason as well as to keep pressure drop low, the catalyst beds are thin and wide.

I. Paraffins to Olefinic Hydrocarbons

A. n-Butane

Large quantities of butylenes are required as intermediates in making butadiene for synthetic rubber. Chromia on γ-, or η-alumina is the common catalyst and typical operating conditions are (*15*, *18*): Temperature 550°–600°C, pressure 0.3–3 atm, LHSV 1–3 $V_l V_c^{-1}$ hr^{-1}, and process period 10–100 minutes. Conversion per pass is in the range of 30–40%, with ultimate yields of 70–80%. Around 5% of the butane converted goes to coke on the catalyst and is burned.

It is not only usual to preheat the feed to supply heat of reaction. The catalyst is also mixed with an inert material which acts a heat sink so that the heat liberated during regeneration is kept in the catalyst bed and used for the process.

Reaction vessels are made of high chrome steels. Chrome–nickel stainless steels may become catalytic and cause excess coke deposition.

With catalysts bringing the reaction mixture close to equilibrium, numerous tricks have been tried to displace the equilibrium in a favorable direction. Operating at reduced pressure helps but is expensive. Adding reagents to react with the hydrogen produced (e.g., CO_2 or SO_2) has not proved practical.

B. Isobutane

Isobutane can be thermally cracked to give approximately equal yields of propylene and isobutylene and at this writing the process is being employed to utilize excess isobutane from the hydrocracking of petroleum. Isobutane dehydrogenates catalytically to isobutylene under the same conditions and with the same catalysts used for *n*-butane. So far as the author is aware, the process is not being used commercially as of this writing.

C. Ethane

Ethane dehydrogenates thermally to ethylene and acetylene so is not considered for catalytic dehydrogenation.

D. Propane

Propane dehydrogenates to propylene with the chromia–alumina catalysts mentioned above. At temperatures around 600°C the equilibrium yield is too small to make the process economic. Propylene from thermal cracking or catalytic cracking of other petroleum fractions is cheaper.

E. n-Pentane

n-Pentane dehydrogenates to a mixture of *n*-pentenes when the chromia–alumina catalysts are used at 500°–550°C. So far as the author is aware, this process is not used commercially.

F. Isopentane

Isopentane dehydrogenates similarly to give a mixture of isopentenes. The process is used as one step towards isoprene for synthetic "natural" rubber. The technology is similar and patterned after the dehydrogenation of *n*-butane to butylenes to butadiene (see II A below).

G. n-Hexane and Higher n-Paraffins

These hydrocarbons cyclicize with the chromia–alumina catalysts and go all the way to aromatic hydrocarbons although at high space velocities, small yields of the olefin are found apparently as intermediates. *n*-Hexane gives benzene; *n*-heptane, toluene; *n*-octane, xylenes plus ethylbenzene; etc. Typical conditions are 500°C, atmospheric pressure, and LHSV 1–4 $V_l V_c^{-1}$ hr^{-1}. The process periods are 1–6 hours, after which the catalyst is regenerated by burning off the coke deposit, and then reused. The author is not aware that this process is being used commercially at this writing.

If molybdena–alumina is used as the catalyst in place of chromia–alumina, it produces aromatics and also isomerizes paraffins. However, it has been found that by using a partially deactivated catalyst in the presence of steam and keeping the total conversion per pass low, i.e., 12% the conversion, to monoolefin was 8–9% with a selectivity of 60–65% (*21*, *23*). The conditions with *n*-dodecane were 470°C, 4:1 steam:hydrocarbon mole ratio, and atmospheric pressure.

A process in which the higher *n*-paraffins are dehydrogenated to *n*-olefins has been offered commercially (*4*, *7*) but as of this writing the nature and properties of the catalyst have not been made public.

H. Branched Chain Paraffins

Interestingly, branched chain paraffins containing less than six carbon atoms in the longest chain but with a total of six or more carbon atoms, can be converted into aromatic hydrocarbons by chromia–alumina catalysts containing alkali, e.g., "K_2O" (*19*). (If alumina without alkali were used, it might be argued that the alumina was acidic enough to isomerize the paraffin to one with a chain containing six or more carbon atoms before aromatization. The presence of the alkali seems to negate this argument.)

2,2,4-Trimethylpentane gave *p*-xylene (*9*, *12*, *20*), while 2,3,4-trimethylpentane gave all three xylenes plus ethylbenzene. 2,2,3-Trimethylpentane gave mostly *m*-xylene and toluene (*9*).

Although 1,1,3-trimethylcyclopentane is a logical intermediate with 2,2,4-trimethylpentane, at best only traces were isolated with chromia–alumina. Further, 1,1,3-trimethylcyclopentane over the catalyst gave mixed xylenes instead of just *p*-xylene (*9*). In contrast, it has been shown that with base promoted Pt–alumina, 2,2,4-trimethylpentane gave considerable yields of 1,1,3-trimethylcyclopentane in the presence of H_2 at 425°C. In addition, a 40:60 mixture of *p*- and *m*-xylene was produced instead of just *p*-xylene, indicating the difference between Pt–alumina and chromia–alumina (*17*).

TABLE I

DEHYDROGENATION OF TRIMETHYLPENTANES BY PT–ALUMINA[a]

Trimethylpentane	Trimethylcyclopentane	Xylenes formed (%)
2,2,4–	1,1,3–	40 *p*–, 60 *m*–
2,2,3–	1,1,2–	22 *o*–, 78 *m*–
2,3,3–	1,1,2–	22 *o*–, 78 *m*–
2,3,4–	1,2,3–	43 *p*–, 43 *o*–, 14 *m*–

[a] See reference (*17*).

Table I gives the starting trimethylpentane, the cyclopentane intermediate and the xylenes formed in a series of tests with the Pt–alumina catalyst (*17*).

I. Chromia–Alumina Catalyst Poisons

Chromia–alumina catalysts are remarkably insensitive to poisons which might be present in the starting material. Water vapor does act as a temporary poison, so the input should be dry. The system acts as if water competes with the hydrocarbon for active sites on the catalyst. This means that oxidative dehydrogenation, if and when it occurs, probably will not be based on chromia–alumina catalysts, since water is formed in the reaction.

This is also the reason that attempts to use the catalyst in the presence of steam to remove the coke deposit as CO and H_2 or CO_2 and H_2 have failed.

J. Regeneration of Chromia–Alumina Catalysts

The chromia–alumina catalysts in dehydrogenation are used in relatively thin fixed beds of granules to reduce pressure drop. The coke buildup occurs fairly fast, so that regeneration must occur every few minutes up to every hour. The catalyst may be mixed with an inert "heat sink" to carry heat from the regeneration over to the endothermic process, in which case air may be used to remove the coke. Otherwise, recirculated flue gas containing 2–3% oxygen is used and the maximum permissible temperatures are about 650°C. Higher temperatures convert the Cr_2O_3 to the pink or purple form (inactive) and the γ-alumina to α. As has been mentioned in Chapter 2, little or no CO is formed.

1–2% K_2O (probably as K_2CO_3) in the catalyst reduces the coke yield but makes the catalyst somewhat more sensitive to excessive temperature during regeneration.

K. Suppliers of Proven Catalysts

Catalysts Division of Chemetron Corp.
Girdler G–41: 7.5% Cr on alumina.

Davison Chemical Division of W. R. Grace & Co.
Grade 909: chromia–alumina.

Harshaw Chemical Company
Cr–0101 T: 12% Cr_2O_3 plus 2% MgO on activated alumina, ⅛ inch tablets. The ABD is 1.0, the surface area 45 m^2gm^{-1}, and the pore volume 0.31 cm^3gm^{-1}.
Cr–0103 T: 12% Cr_2O_3 on activated alumina; ⅛ inch tablets. The ABD is 1.0, the surface area 63 m^2gm^{-1}, and the pore volume 0.35 cm^3gm^{-1}.
Cr–0104 T: 5% Cr_2O_3 on activated alumina, ⅛ inch tablets. The ABD is 1.1, the surface area 80 m^2gm^{-1}, and the pore volume 0.36 cm^3gm^{-1}.
Cr–0105 T: 9% Cr_2O_3 plus 1.5% K_2O on activated alumina, ⅛ inch tablets. The ABD is 1.1, the surface area 67 m^2gm^{-1}, and the pore volume 0.34 cm^3gm^{-1}.
Cr–0205 T: 19% Cr_2O_3 on activated alumina (specially designed for butane to butadiene with butylenes as intermediates) as 5/32 inch tablets (Spec. 102). The ABD is 1.0, the surface area 60 m^2gm^{-1}, and the pore volume 0.29 cm^3gm^{-1}.
Cr–0211 T: 19% Cr_2O_3 on alumina, 5/32 inch tablets. The ABD is 1.16, the surface area 62 m^2gm^{-1}, and the pore volume 0.3 cm^3gm^{-1}.
Cr–0215 T: 19% Cr_2O_3 on alumina: 5/32 inch tablets. The ABD is 1.2, the surface area 65 m^2gm^{-1}, and the pore volume 0.3 cm^3gm^{-1}.
Cr–0304 T: 33% Cr_2O_3 on silicated alumina, ⅛ inch tablets. The ABD is 1.2, the surface area 120 m^2gm^{-1}, and the pore volume 0.26 cm^3gm^{-1}.
Cr–1404 P: 19% Cr_2O_3 on alumina as microspheres. The ABD is 0.9, the surface area 80–100 m^2gm^{-1}, and the pore volume 0.50 cm^3gm^{-1}.
Cr–1404 T: the ⅛ inch tabletted form of Cr–1404 P. The ABD is 1.2, the surface area 80–100 m^2gm^{-1}, and the pore volume 0.38 cm^3gm^{-1}.
Mo–0502 T: 10% MoO_3 on activated alumina, ⅛ inch tablets. The ABD is 1, the surface area is 64 m^2gm^{-1}, and the pore volume 0.40 cm^3gm^{-1}.
Mo–1201 T: 10% MoO_3 on high activity alumina, ⅛ inch tablets. The ABD 0.85, the surface area 160 m^2gm^{-1}, and the pore volume 0.36 cm^3gm^{-1}.

Houdry Process and Chemical Co.
Chromia–alumina Series A Grade 25Z and Series C.

Universal Oil Products Co.
J–6 chromia–alumina.

II. Olefins to Diolefinic Hydrocarbons

A. Butylenes to Butadiene

The same chromia–alumina catalysts which dehydrogenate butane to butylenes also dehydrogenate butylenes to butadiene. Thus, by separating the hydrogen and recycling the primary mixture of butane and butylenes, a mixture of butane, butadiene, and butylenes is produced. The butadiene is separated, and the unconverted butylenes and butane recycled along with added butane. Preferred conditions are (*5*, *15*, *18*): Temperature 550°–650°C, 15–20 cm Hg absolute pressure, LHSV 1–3 $V_l V_c^{-1}$ hr^{-1}, and process periods of 7–15 minutes. The preferred catalyst is 18–20% chromia on active (γ or η)alumina.

Dehydrogenation of butylenes to make butadiene is meeting stiff competition from oxidative dehydrogenation of butylenes. This is discussed in Chapter 20.

Both poison and regeneration considerations are the same as for paraffin to olefin dehydrogenation with chromia–alumina catalysts.

Catalysts have been developed that can dehydrogenate olefins to diolefins in the presence of steam. (They are not active for converting paraffins to olefins.) Three major benefits favor these catalysts (*8*, *14*, *15*, *22*):

(*a*) Low partial pressures are obtained with systems operating at total pressures of atmospheric or above.

(*b*) The steam reacts with the coke deposits extending the process periods greatly.

(*c*) Highly preheated steam helps carry the needed heat into the reactor.

Typical operating conditions are: Temperature 590°–690°C, C_4H_8 GHSV 150–800 $V_g V_c^{-1}$ hr^{-1}, steam:C_4H_8 ratio 8:1 to 12:1 moles, conversion per pass 25–30%, and selectivity to C_4H_6 65–80%.

The high reaction temperatures are usually obtained by preheating the steam to a higher temperature (e.g., 725°C) than the butylenes (600°C) and then mixing just before the catalyst bed. The products are quickly quenched with steam or water to prevent secondary reaction. Butadiene is selectively removed from the C_4 fraction and the rest of the butylenes recycled.

When coke accumulates on the catalyst, the steam regenerates the catalyst while the hydrocarbon feed is stopped.

This process too is feeling the pressure from the oxidative dehydrogenation of butylenes to produce butadiene (see Chapter 20).

B. Pentenes

Pentenes give lower yields of the corresponding dienes when using catalysts developed for dehydrogenating butylenes. Chromia–alumina with isopentene gave up to 67% selectivity to isoprene at a 70% conversion level. With the catalysts active in the presence of steam, isopentene gave 66% selectivity to isoprene at 42% conversion. In contrast, 2-pentene gave only 33% selectivity to piperylene.

C. Suppliers of Proven Catalysts

Catalysts Division of Chemetron Corp.
G–48: an iron base catalyst.

Dow Chemical Co.
Dow Type B Dehydrogenation Catalyst: a nickel–calcium phosphate $[Ca_8Ni(PO_4)_6]$ stabilized with chromium oxide (*8*). High selectivity (80–90%) to butadiene at butylene conversion levels of 40–50% per pass are indicated for this catalyst. Large steam:butylene ratios (18:1 or larger) plus periodic regeneration with steam + air are apparently necessary to obtain this high selectivity (*14*, *22*).

The Ammonia Division of Shell Chemical Co.
(The following two catalysts are for use with steam.)
Shell 105: 87% Fe_2O_3, 10% K_2O, and 3% Cr_2O_3.
Shell 205: 70% Fe_2O_3, 27% K_2O, and 3% Cr_2O_3.

It is well known that iron, cobalt, and nickel metals can be potent catalysts for the complete decomposition of hydrocarbons into coke and hydrogen. Here are catalysts containing iron or nickel compounds but which can be used under conditions that prevent their reduction to the undesirably catalytic metals. The large excess of steam or the combination of steam and the other catalyst components prevents it. Hence, excess coke formation may occur if one attempts to use these catalysts with hydrocarbons alone under these conditions.

III. Ethylbenzene to Styrene

Ethylbenzene yields styrene on pyrolysis, but the selectivity is poor. Catalytically, it dehydrogenates a little more easily than butylene and the same catalysts are useful. Thus, chromia–alumina will work, but reduced pressure and catalyst regeneration are necessary. As a result, the iron oxide based catalysts discussed in the previous section are usually used. The temperature range is 590–650°C. The ethylbenzene is mixed with preheated steam just prior to entering the reactor (isothermal or adiabatic reactor). The quantity of steam is 1–3 kg per kg of ethylbenzene. Older single stage units obtain 85–90 mole % selectivity while more modern multistage reactors achieve 90–91% at conversions of 55–60%.

2-Methyl-5-ethylpyridine dehydrogenates to 2-methyl-5-vinylpyridine under quite similar conditions.

A. Suppliers of Proven Catalysts

See Section I–K, Chapter 6 for the chromia–alumina catalysts and Section II–C, Chapter 6 for the iron and nickel base catalysts.

Catalysts Division of Chemetron Corp. supplies both Girdler G–48 and G–84 for ethylbenzene dehydrogenation.

IV. Cycloparaffins

Cyclopentane at 500°C gives only modest yields of cyclopentene with the dehydrogenation catalysts (chromia–alumina) and coke formation is relatively severe. The author does not know a good catalyst for this reaction.

As has already been mentioned, the homologs of cyclopentane isomerize and dehydrogenate to aromatic hydrocarbons. Using an acidic alumina as the base aids the isomerization.

The cyclohexanes dehydrogenate readily to aromatic hydrocarbons in the presence of chromia–alumina, molybdena–alumina, or Pt–alumina catalysts. If one is seeking a nearly quantitative laboratory procedure, these catalysts are not always suitable. Because the acidity of the alumina may isomerize some of the cyclohexanes to cyclopentane homologs, the reaction is not quantitative. Activated carbon is a neutral support and 5% Pt or 5% Pd on activated carbon is often used at 200°–300°C and atmospheric pressure. It should be noted that nickel catalysts are not completely reliable for this dehydrogenation because sometimes they decompose part of the hydrocarbon to carbon and hydrogen or methane and hydrogen.

Under different circumstances, the cyclohexanes can be used as hydrogen donors as they dehydrogenate to aromatics. Decalin, particularly, has been used both as a solvent and hydrogen source in research on the hydrogenation of coal (*24*). The decalin was converted to tetralin and naphthalene in the process.

The writer does not know of any commercial process using this reaction. If faced with choosing a catalyst for the reaction, he would select a catalyst that would be expected to hydrogenate the recipient molecule.

A. Suppliers of Proven Catalysts

See Chapter 13.

V. Cycloolefins

Cyclopentene at 500°C dehydrogenates to cyclopentadiene with chromia–alumina but the yields are poor due to the large coke formation.

Cyclohexene dehydrogenates readily to benzene over chromia–alumina, molybdena–alumina, and Pt–alumina catalysts at 400°–500°C. No significant amount of cyclohexadiene is formed.

It has been known for many years that cyclohexene would readily convert to cyclohexane and benzene just by refluxing with an active Pd catalyst (*26*). This is true of the cyclohexene homologs as well. It also is true of other hydrogenation catalysts. 5% Pt, 5% Pd on active charcoal, or nickel on kieselguhr are satisfactory. Interestingly, the cycloolefin can act as a hydrogen donor to other molecules as it loses hydrogen in becoming an aromatic. In this respect, it more readily donates its hydrogen than the cyclohexanes referred to in an earlier section. It is only necessary to dissolve the substance to be hydrogenated in cyclohexene and reflux in the presence of the active hydrogenation catalyst. This transfer of hydrogen need not be just to hydrocarbons; it can be unsaturated esters, nitro compounds, etc. Obviously, the thermodynamics must favor the hydrogen recipient; otherwise, the hydrogen will go to the cyclohexene to form cyclohexane.

The reaction can be relatively slow in that refluxing for 50–100 hours may be required. The user should not reflux just for an hour or two and decide his system is not working.

Cyclohexene and its homologs dehydrogenate to the corresponding aromatic hydrocarbons with the Pt or Pd catalysts, but this requires higher temperatures than the hydrogen disproportionation reaction. In fact the temperature is in the same range as that required for the cyclohexanes, i.e., 200°–300°C and atmospheric pressure. As with the cyclohexanes the author prefers 5% Pt or 5% Pd on activated carbon as catalysts for laboratory use.

It must be kept in mind that the cyclohexenes peroxidize readily and should be redistilled just before use. A generous residue should be left in the still pot lest the concentrated peroxides explode.

A. Suppliers of Proven Catalysts

See Chapter 13 for nickel, Pd, and Pt catalysts.

VI. Alcohols to Aldehydes and Ketones

The catalytic dehydrogenation of primary and secondary alcohols to give aldehydes or ketones has been known for many years. Ethyl alcohol gives acetaldehyde; isopropyl alcohol, acetone; *sec*-butyl alcohol, methyl ethyl ketone; etc. Methyl alcohol is usually oxidized to formaldehyde since as shown below, dehydrogenation gives CO plus H_2. The most widely used catalysts are based on metallic copper. This may be alone, as a brass, or in combinations such as copper chromite. Zinc oxide has been used in a number of kinetic studies (*16*).

Conditions are 250°–400°C and substantially atmospheric pressure. The space velocity varies with the alcohol and catalyst but is usually in the range of 2–5 $V_l V_c^{-1}$ hr^{-1}. The reaction is endothermic and conversion is rarely over 50% per pass.

Silver catalysts are also effective but rarely justify the incremental cost.

Note that η- or γ-alumina is not usually a suitable carrier for the vapor phase alcohol catalysts, for they induce dehydration to the olefin. Other dehydration carriers are also ordinarily excluded.

The cyclohexyl alcohols present a special case. The copper based catalysts or Ru on activated carbon will dehydrogenate to the cyclohexanone. On the other hand, Pt on activated carbon will dehydrogenate to the phenol. The temperature is 325°–400°C, LHSV 0.6–1.25 $V_l V_c^{-1}$ hr^{-1}, and pressure atmospheric (*6*).

A process for making phenol employs a dehydrogenation–hydrogenation sequence. Benzene is hydrogenated to cyclohexane which in turn is oxidized with air (cobalt acetate catalyst, 170°–180°C, 30 atm pressure) to a mixture of cyclohexanol + cyclohexanone that is dehydrogenated over a nickel catalyst to give phenol (*3*).

A prereduced (375°C) 8.7% Ni + 7.1% Sn on silica gel gave better activity and selectivity to phenol than 12% Ni on silica gel (*25*). The author does not know a commercial supplier for this catalyst.

Fairly high boiling liquid alcohols can be dehydrogenated in the liquid phase in the presence of the powdered dehydrogenation catalysts, e.g., active Raney nickel. It is possible to aid the reaction by bubbling air through the mixture to oxidize the hydrogen as it is formed (*13*). The secondary alcohols are especially susceptible to this type of operation. It is also possible to suspend the catalyst in a high boiling solvent for the alcohol, heat the solvent to dehydrogenation temperature, and gradually introduce the alcohol while distilling off the ketone (*1*).

Ordinarily no attempt is made to regenerate catalysts which are used as suspended powders in reactions such as those described above.

A. Suppliers of Proven Catalysts

See Chapter 13 for Pt, Ru, and Raney nickel catalysts and Chapter 15 for copper chromite catalysts. The copper chromite catalysts must be reduced in hydrogen at 200°C–300°C before use as dehydrogenation catalysts. Usually the catalysts without BaO are preferred for this type dehydrogenation.

Harshaw Chemical Company

Cu–2501 G: an alkali promoted catalyst containing 6% Cu as carbonate on a silica support, 4–10 mesh granules. The ABD is 0.5, the surface area 4 m^2gm^{-1}, and the pore volume 0.61 cm^3gm^{-1}.

VII. Methanol to Hydrogen and Carbon Monoxide

Zinc chromite is a classical catalyst in that it will catalyze the synthesis or the decomposition (dehydrogenation) of methanol. As a result, the catalyst is used to decompose methanol to make CO and H_2. Where these are needed in relatively small quantities the operation is cheap enough to justify the raw material cost.

The conditions are: Temperature 350°–450°C, atmospheric pressure, LHSV 1–5 $V_1V_c^{-1}$ hr^{-1}.

To some extent, zinc chromite has been considered for the related reaction:

$$CH_3OH + H_2O \rightarrow 3\ H_2 + CO_2$$

where limited quantities of hydrogen are desired, e.g., for filling balloons during World War II. However, $Mn(OH)_2$ deposited on silica gel is so much more reactive that zinc chromite was displaced. The catalyst operates in boiling methanol–water under pressure, i.e., 130°C and 15 atm.

A. Supplier of Proven Catalysts

Harshaw Chemical Company

Zn–0302 T: 74% ZnO plus 23% chromium oxide as hexavalent Cr, 3/16, 1/4 and 3/8 inch tablets. The catalyst should be reduced in hydrogen before use.

Zn–0308 T: this is the reduced form of Zn–0302 T, 1/4 inch tablets.

Mn–0501 G: 2% Mn in oxide form on silica, 3–8 mesh granules. The ABD is 0.5, the surface area 258 m^2gm^{-1}, and the pore volume 0.97 cm^3gm^{-1}.

VIII. Ammonia to Hydrogen and Nitrogen

This reaction may not ordinarily be thought of as a dehydrogenation but it has all the characteristics. It is used to generate small quantities of reducing gas, e.g., for metallurgical use. The temperature of decomposition is 875°–1000°C, atmospheric pressure, and GHSV 2000 $V_gV_c^{-1}$ hr^{-1}. Catalysts of the type used for steam reforming of natural gas, i.e., Ni on refractory supports, such as mullite, or iron oxide on similar supports, are used.

The decomposition of ammonia is used also in small thrusters on space vehicles (*10*). Using radioactive promethium sesquioxide to provide heat, the small reactor containing a ruthenium catalyst is permanently heated to 1000°C. As thrust is needed, small slugs of ammonia are passed over the catalyst at a WHSV on the order of 100,000 $W_r W_c^{-1}$ hr^{-1}. At such high velocities, the catalyst particles have to be fitted in the reactor so they will not be blown out.

A. Suppliers of Proven Catalysts

Catalysts Division of the Chemetron Corp.
- Girdler G–29 and G–56: nickel on inert, refractory supports.
- G–47: an iron oxide catalyst.

Both Engelhard Industries and Matthey Bishop, Inc. supply ruthenium catalysts.

Harshaw Chemical Company
- Ni–0901 S: 5% Ni present as oxide on inert alumina–silica spheres ⅜, ½, ¾ or 1 inch in diameter. Bulk density is 1.2 and surface area is 1 m^2gm^{-1}.
- Fe–1401 S: 5% Fe_2O_3 on inert alumina spheres ⅜ inch in diameter. Bulk density is 1.2, the surface area 1 m^2gm^{-1}, and the pore volume 0.06 cm^3gm^{-1}.

IX. Methane and Ammonia to Hydrogen Cyanide

Heating a mixture of natural gas + ammonia at 1150°–1200°C produces HCN and hydrogen. The reaction is strongly endothermic. The catalyst is Pt deposited on the inside of alundum (α-alumina) tubes which serve as reactors and are heated externally. The yield based on NH_3 is 80–85%, and 85–90% based on CH_4. At first the catalyst carbonizes fairly rapidly and must be regenerated. On each successive regeneration the activity and life improve. After 400 hours and 3–5 regenerations, no further coke was formed and there was no rapid deactivation. The author knows of no commercial use of the process. Rather it is of special interest to the catalyst chemist because of the changes taking place in the catalyst. Examination of the catalyst which no longer carbonized showed Pt_3Al and Pt_5Al as well as AlN. Apparently, some of the α-alumina was reduced to metallic Al which alloyed with the Pt. Dispersing finely powdered alumina in with the platinum compounds during the coating of the tubes was said to shorten the initial "activation" period to a few hours, presumably by supplying an alumina which was more readily reduced so it could form the alloy (*11*).

If the reader is giving serious consideration to this type of HCN process, he should take into consideration the competing electrothermal process which is noncatalytic. A fluidized bed of coke is electrically heated to >1400°C and natural gas or other hydrocarbons plus ammonia passed in to produce HCN and hydrogen (*2*).

REFERENCES

1. Anonymous, *Hydrocarbon Process. Petrol. Refiner* **42,** No. 11, 191 (1963).
2. Anonymous, *Hydrocarbon Process. Petrol. Refiner* **42,** No. 11, 183 (1963).
3. Anonymous, *Hydrocarbon Process. Petrol. Refiner* **44,** No. 11, 237 (1965).
4. Anonymous, *Hydrocarbon Process.* **46,** No. 11, 197 (1967).
5. Anonymous, *Hydrocarbon Process.* **47,** No. 9, 230 (1968).
6. Belgian Patent 616,103 (Oct. 8, 1962); 621,338 (Feb. 13, 1963); 623,725 (Apr. 17, 1963).
7. Bloch, H. S., *Oil Gas J.* **65,** No. 3, 80 (1967).
8. Britton, E. C., Dietzler, A. J., and Noddings, C. R., *Ind. Eng. Chem.* **43,** 2871 (1951).
9. Cannings, F. R., Fisher, A., Ford, J. F., Holmes, P. D., and Smith, R. S., *Chem. Ind.* (*London*) p. 228 (1960).
10. Cohn, J. G., *Engelhard Ind. Tech. Bull.* **9,** No. 1, 5 (1968).
11. Endter, F., Am. Chem. Soc. 147th Meeting, Philadelphia, Penna., 1964. Div. Ind. Eng. Chem., Paper No. 28.
12. Herington, E. F. G., and Rideal, E. K., *Proc. Roy. Soc.* (*London*) A**184,** 434 (1945).
13. Heyns, K., and Paulsen, H., *Angew. Chem.* **69,** 600 (1957).
14. Jones, J., *Dow Diamond* Spring, pp. 24–27 (1962).
15. Kearby, K. K., *Catalysis* **3,** 453 (1955).
16. Kolboe, S., *J. Catalysis* **13,** 193 (1969).
17. Lester, G. R., *J. Catalysis* **13,** 187 (1969).
18. McKetta, J. J., Jr., *Advan. Petrol. Chem. Refining* **4,** 450 (1961).
19. Payne, J. W., Evans, L. P., Bergstrom, E. V., and Bowles, V. O., *Petrol. Refiner* **31,** No. 5, 117 (1952).
20. Pines, H., and Csicsery, S. M., *J. Catalysis* **1,** 313 (1962).
21. Pitkethly, R. C., and Steiner, H., *Trans. Faraday Soc.* **35,** 979 (1939).
22. Reilly, P. M., *Chem. Can.* March, pp. 25–29 (1953).
23. Roth, J. F., Abell, J. B., and Schaffer, A. R., *Ind. Eng. Chem. Prod. Res. Develop.* **7,** 255 (1968).
24. Storch, H. H., *in* "Chemistry of Coal Utilization" (H. H. Lowry, ed.), Vol. 2, p. 1750. Wiley, New York, 1945.
25. Swift, H. E., and Bozik, J. E., *Am. Chem. Soc. Div. Petrol. Chem. Preprints* **13,** No.3, 173 (1968).
26. Zelinsky, N. D., and Pavlov, G. S., *Ber. Deut. Chem. Ges.* **66,** 1420 (1933).

CHAPTER 7

CATALYTIC REFORMING OF NAPHTHAS

The gasoline which occurs in crude oil is too poor in antiknock quality to be used alone as fuel in the modern automobile engine. The petroleum industry uses catalytic reforming to improve gasoline antiknock quality. Several broad types of hydrocarbon reactions occur although not all of them are significant in all the processes:

(*a*) Dehydrogenation of cyclohexanes to aromatic hydrocarbons.

(*b*) Isomerization of cyclopentane homologs to cyclohexanes, followed by dehydrogenation to aromatics.

(*c*) Isomerization of paraffin hydrocarbons. (This is discussed in Chapter 3).

(*d*) Cyclization of paraffins followed by dehydrogenation to aromatics.

(*e*) Hydrocracking of poor antiknock paraffins to gases, thus concentrating the good antiknock aromatics in the liquid.

(*f*) Hydrogenation of coke formers on the catalyst.

There are several good review summaries on catalytic reforming giving many more details than can be given here (*5, 7, 9–11, 16, 17*).

I. Platinum Reforming

This is the reforming process having the widest current use. It brought the significance of dual function catalysts into both prominence and importance. These catalysts, which are 0.3–1.0% Pt well dispersed on an acidic alumina (γ or η), have dehydrogenation sites on the Pt and isomerization sites on the alumina. Cl^-, F^-, or both may be added to the alumina to get the desired acidity. In some respects these two functions seem to operate completely independently, e.g., methylcyclohexane dehydrogenates on the Pt as if the alumina were not there. The cyclopentane homologs isomerize on the alumina almost as if the Pt were not there, yet the two must cooperate in some way in keeping the catalyst clean. It is clear that the coke tends to form primarily on the acidic alumina, but it is the Pt which is the catalyst for hydrogenating the coke or coke formers. Just how this takes place is one of the intriguing problems of this process and hydrocracking.

In catalytic reforming, the dehydrogenation of the cyclohexanes is both fast and strongly endothermic. The aromatization of paraffins is slower but also endothermic. For this reason, platinum reforming consists of several fixed bed adiabatic reactors in series with reheating in between to supply the heat of reaction.

Isomerization of the cyclopentane homologs is slow and nearly isothermal. This means that sufficient catalyst volume must be provided to take care of this slow reaction if these hydrocarbons are to be utilized fully.

In the dehydrogenation reactions one would expect the process to be conducted at or near atmospheric pressure. However, under these conditions the catalyst cokes rapidly. By increasing the pressure and recirculating hydrogen, the coking is slowed down. At pressures around 10 atm, the catalyst can be operated for a few days before regenerating it. By operating at higher pressures, up to 40 atm, the catalyst can be kept clean enough so that it can be operated several months without regeneration. Both types of operation are used, so it is convenient to discuss them separately as "low" and "high" pressure reforming.

A. Low Pressure Platinum Reforming

This process is used when the maximum antiknock quality product is wanted. The lower pressure used favors the aromatics at equilibrium and so tends to leave less unconverted cycloparaffins in the product.

Three or four fixed bed catalyst cases, all of the same size, are connected in series each preceded by a fired heater. An additional reactor, called a "swing" reactor, is used as follows. All reactors are fitted with valves so that, as the catalyst in any one becomes fouled, it can be cut out of the circuit, becoming the "swing" reactor, and be regenerated. Meanwhile, the freshly regenerated catalyst in the "swing" reactor is put into the series so it can be used to best advantage. In one preferred operation, the catalyst with the longest on stream time since regeneration is used as the first reactor in the series. The freshly regenerated catalyst is the last in the series. Thus, the low concentration of reactant remaining gets acted upon by the most active catalyst.

Typical operating conditions are as follows (*1*): Temperature 480°–520°C; pressure 13–20 atm; H_2 Circulation 2–5 moles per mole of hydrocarbon; LHSV 1–3 $V_l V_c^{-1}$ hr^{-1}. Space velocity and hydrogen circulation are adjusted to meet the antiknock quality requirements. Temperature is adjusted very slowly upwards over a period of months as the catalyst loading gradually loses activity.

Each individual reactor is on stream several days before it is regenerated.

B. High Pressure Platinum Reforming

Like low pressure reforming, high pressure reforming operates with three or four fixed bed catalyst cases in series, each preceded by a fired heater. However, all the cases are not the same size. The first is the smallest and the last is the largest, with the last case containing as much catalyst as all the preceding ones. For example, the LHSV in the first case in the series may be 18, the second 10, and the third 1. This arrangement recognizes that the initial dehydrogenation is fast and takes place mainly in the first case. The larger cases take care of the much slower isomerization and aromatization reactions. Sometimes the final case is operated to hydrocrack poor-antiknock paraffins.

Typical operating conditions are (*2*): Temperature 470–530°C; pressure 15–40 atm; LHSV 1–3 $V_l V_c^{-1}$ hr^{-1}; H_2 circulation 3–10 moles per mole of hydrocarbon.

With new or freshly regenerated catalyst, the inlet temperatures are at the low end of the range given. Since the catalyst is operated for months without regeneration, it gradually loses activity. To maintain both production rate and antiknock quality, the inlet temperature is slowly raised to compensate. Because regeneration rarely brings the catalyst back to "fresh" activity, the length of time between regenerations gets shorter and shorter until the catalyst must be replaced. The ultimate life of the catalyst is related to the severity of the operation, i.e., the antiknock quality of the product; the better the antiknock quality, the shorter the life. Catalyst life is in the range of 70,000–140,000 liters per kilogram of catalyst.

It is not essential that all reactors operate at the same temperature or hydrogen : hydrocarbon ratio. For example, the hydrogen : hydrocarbon mole ratio may be relatively low (e.g., 2.5 : 1) in the first reactor, where the coking tendency is relatively low, and raised in the final reactor (e.g., to 9 : 1), where hydrocracking and aromatization increase the coking tendency (*15*).

During the long term use of these catalysts, the alumina base loses some surface area and activity. Mainly, however, the small Pt crystallites aggregate, thus decreasing the Pt surface area and, consequently, the hydrogenation–dehydrogenation function. The latter change is usually the fastest, resulting in a change in the dehydrogenating to acidic activity ratio. This in turn leads to more cracking and a decline in gasoline yield, with an increase in gas yield. The more severe the operating conditions, the greater the tendency for the Pt crystallites to aggregate.

Pt aggregation is also related to Pt concentration. Thus low pressure reforming can operate with catalysts containing a relatively low Pt concentration (e.g., 0.3–0.4%) and yet the conditions can be regarded as severe (*6*).

The growth of the Pt crystallites can also be inhibited by adding rhenium to the catalyst (*14*). Apparently the Re forms an alloy with the Pt that is more stable than the Pt. The amount of Pt in the catalyst is not reduced significantly, and, while the amount of Re can vary, it is roughly equal to the Pt content.

Advantage is being taken of these more stable catalysts in several ways. For example, in high pressure reforming, the length of the period between regenerations can be extended greatly, the severity can be increased to produce a further improved antiknock quality product, or the pressure can be reduced into the low pressure reforming range (15 atm or less) to get the benefits of low pressure operation while retaining the 6–9 months between regenerations (*13*, *21*).

C. Platinum Reforming for Petrochemicals

Most reformers are operated to produce motor fuel. A few are operated to produce benzene, toluene, and mixed xylenes. Either the high or the low pressure operation can be used. The main difference from ordinary reforming is the fractionation of the feed into a narrow boiling range to facilitate later separation of the aromatic hydrocarbons from the paraffins.

D. Platinum Reforming Plus Selective Cracking

The antiknock quality of the reforming product ("reformate") is controlled by the proportion of poor-antiknock nonaromatic material left in it, mostly paraffin hydrocarbons. Since the paraffins are more susceptible to thermal cracking than the aromatics, it is possible to remove them as gases by thermally cracking them (*3*). In an alternative but catalytic method, a shape-selective dual function catalyst permits selective hydrocracking and removal of the poorest antiknock hydrocarbons, the *n*-paraffins (*4*). Apparently, the cracking function is based on a zeolite whose pores admit only *n*-paraffins, e.g., Linde 5A.

E. Poisons in Platinum Reforming

Arsenic, copper, or lead compounds act as permanent poisons.

It will be recalled that the alumina of the catalyst is an acidic alumina.

The natural acidity is usually augmented by some Cl^-, F^-, or both, with halide contents averaging around 0.5–1.0%. Basic nitrogen compounds in the reactant will neutralize some of this acidity and act as a temporary poison for the isomerizing and cracking function. Mostly the N content of the feed is kept below 0.5 ppm.

In a similar way, sulfur compounds act as temporary poisons for the Pt. It is possible to operate a Pt–alumina reformer with about 0.1% S in the feed but the yield–antiknock quality relation of the operation is unfavorable. Unless some way is provided to remove H_2S from the recirculating H_2, the H_2S builds up and the partial pressure of H_2S in the system is greater than would be expected from the S content of the feed alone. Because of this, hydrotreaters (hydrodesulfurizers) precede the reformer to reduce N to <0.5 ppm and S to <20 ppm, and the resulting NH_3 and H_2S are removed before going to the reformer.

Organic chlorides in the feed react with the alumina, increase the acid function, and increase hydrocracking, thus decreasing the gasoline yield. Sometimes the acid function may need augmenting, so small quantities of organic chlorides, e.g., ethylene dichloride, are added to the feed. Organic chlorides may get into the feed as a result of chlorinated solvents being used in the oil fields for controlling paraffin deposits in pipe lines, tanks, etc. (*22*). Since these solvents concentrate in the feed to the reformer, because of their boiling range, they can damage the catalyst and cause expensive corrosion through the HCl liberated.

Water in the feed and in the recycle hydrogen is avoided by many users and they limit it to 5–10 ppm H_2O. They claim that water tends to hydrolyze Cl^- and remove it from the catalyst as HCl which corrodes equipment and tends to decrease the hydrocracking and isomerizing function. However, other operators find water gives an increase in hydrocracking, at least for a while, and decreases gasoline yield. Actually, the water content of both feed and recycle hydrogen is controlled so both are often dried (*18*). The main difference is that some users prefer to control the water in the 50–80 ppm range instead of the 10 ppm range. The difference is probably related to the specific catalyst being used and if this is important to the reader, he should consult the catalyst supplier.

One of the reforming catalysts uses specially prepared silica–alumina as the isomerizing component of the catalyst instead of chlorided or fluorided alumina. Since halogen hydrolysis cannot take place, this catalyst is less sensitive to water (*8*).

F. Regeneration of Pt–Alumina Reforming Catalysts

Regeneration is done with recirculated flue gas containing about 0.5–1.0% O_2. Ignition is obtained at the lowest practical temperature, around

300°C, and the circulation rate is regulated to control the temperature and keep it below about 450°C. As burning slackens, the O_2 content of the gas is increased to 2%. Some users provide a "soak" after the burning is complete and do this at higher O_2 concentrations, even with straight air and at temperatures of 480°C or higher. Various rationales have been proposed for using this O_2 "soak." The most logical is that it reoxidizes the Pt surface and that it may retard agglomeration of the Pt crystallites and thus conserve Pt surface area and activity.

Water in the recirculating flue gas, formed from burning hydrogen in the catalyst coke, can also cause hydrolysis of the halides on the catalyst and in turn cause corrosion in the plant. Some users dry the recirculating gas to prevent this. Others scrub the circulating gas with water to remove the corrosive gases before it is recirculated, thus preventing corrosion. This latter process does not, of course, prevent hydrolysis of the halide (*18*).

Many users of the high pressure process remove the catalyst from the reactors after regeneration and screen out the fines before returning the catalyst to the reactors. This reduces the pressure drop and can be especially helpful in the first reactor, where the pressure drop can hurt the entire series of reactors.

It is also common practice to "scalp" a catalyst bed as part of the regeneration procedure. This consists in removing scale which has become detached from other parts of the plant and is filtered out of the stream by the catalyst bed. If left on top of the catalyst bed, it increases the pressure drop. The scale may become catalytic during regeneration and catalyze undesired reactions, e.g., coke deposition.

Ultimately, the catalysts lose so much activity that they must be replaced. The catalyst is usually regenerated before removal from the reactor and returned to the catalyst supplier or to the platinum supplier for recovery of the platinum. Some catalyst and some Pt is lost in the loading and later unloading of the reactors. These losses vary widely. Around 1% Pt loss for the loading plus unloading is a reasonable average (*18*).

G. Suppliers of Proven Catalysts

American Cyanamid Company

Aeroform PHF–4: Pt on chlorided gamma alumina in the form of 1/16 and 1/12 inch extrudates.

Aeroform PHF–5: Pt on alumina, a catalyst especially for low pressure reforming.

Chevron Research Company

A reforming catalyst containing platinum and rhenium to licencees of its Rheniforming process.

Engelhard Industries

Sinclair-Baker RD–150: 0.6% Pt on 1/16 inch eta-alumina extrudates.

E–500 series: that produce results like the Pt–Re catalysts.

Houdry Process and Chemical Company furnishes a number of pelleted catalysts

Series A Grade 100S and Series A Grade 200S, which are 0.5% Pt on alumina.

3G, 3H, 3K, and 3L, which contain 0.5–0.75% Pt on alumina.

Universal Oil Products Company

A number of platinum catalysts all supported on gamma-alumina beads. These catalysts are 1/16 or 1/8 inch in diameter and contain from 0.375 to 0.75% Pt plus from 0 to 0.51% F and from 0.23 to 0.90% Cl. They also license a process using a Pt–Re catalyst.

With many millions of dollars invested in Pt in these reformers, it follows that a great deal of research has been done to find less expensive catalysts. Nonplatinum catalysts have been found, but the author knows none at this writing which give the yield–antiknock quality relation provided by the Pt catalysts.

II. Molybdena–Alumina Reforming

The molybdena–alumina reforming process uses the principle of circulating hydrogen at elevated pressure to minimize the formation of carbonaceous material on the catalyst. In a fixed bed operation, the temperatures are around 450°–540°C, pressures 10–20 atm, with recirculating gas to the extent of 2–5 moles per mole hydrocarbon. The LHSV is 0.5 $V_l V_c^{-1}$ hr^{-1} and the process period 2–12 hours (*16*).

The catalyst is regenerated by circulating a heated flue gas–air mixture containing 2–3% O_2 over the catalyst at roughly operating pressure. Catalyst temperatures are controlled to 550°–600°C. Higher temperatures can damage the catalyst by volatilizing some of the molybdena as MoO_3. There is a significant amount of sulfur in the spent catalyst, as well as coke. The sulfur is most likely present as a sulfide of molybdenum and, in a number of cases, approximately 90% of the sulfur in the feed was removed by combining with the catalyst. This was later disposed of as SO_2 during regeneration. This process is primarily of historical interest. It was used during World War II to convert petroleum fractions rich in methylcyclohexane into toluene for the production of TNT.

A version of this reforming process was developed using the fluidized solids technique. It gave 2–4% better yields of product for a given antiknock quality than the fixed bed process. Operating conditions were only slightly different. The temperature was the same, the pressure 12–17 atm, and the H_2 recycle 4–7 moles per mole of hydrocarbon. This reaction, like the other dehydrogenation reactions, is strongly endothermic. To supply this heat, the feed was heated to about 540°C and this furnished about 25% of the heat of reaction. Recycle gas heated separately to 650°C furnished another 55%, and the remaining 20% was carried into the adiabatic reactor by the hot regenerated catalyst (*17*).

Catalyst was removed from the reactor by a suitable valve system and carried to a regenerator where the carbonaceous material was removed by burning with air in a fluidized system under a pressure of about 13–15 atm and 590°–625°C. As in the fixed bed, the catalyst removed sulfur from the feed. It left the process as SO_2 during regeneration.

There is a comparison in the literature of the yields and product quality from the fluid molybdena–alumina reforming process with the low and high pressure Pt–alumina processes which indicates that, for some feeds and product quality levels, the fluidized process was competetive with the low pressure process but not the high pressure one (*20*).

The fluidized molybdena–alumina reforming process received an unfortunate technological setback when a large commercial unit blew up, destroying the unit and causing a serious fire in the refinery. The fact that the explosion was caused by a number of fortuitous circumstances has in no way diminished the suspicion of other potential users of the process. Historically, the process is significant because it represented the first attempt to operate on a large commercial scale a process in which a fluidized catalyst was both used and regenerated at elevated pressures, i.e., 10–15 atm.

A. Suppliers of Proven Catalysts

The name "molybdena"–alumina, it is admitted, does not give an exact description of these catalysts. The molybdenum doubtless changes valence as it goes from process conditions to regenerator conditions. Further, the best catalyst may contain sulfur. So molybdena–alumina expresses what went into the catalyst, and not what the catalyst is.

Harshaw Chemical Company

Mo–0502 T: 10% MoO_3 on activated alumina, ⅛ inch tablets. The ABD is 1, the surface area 64 m^2gm^{-1}, and the pore volume 0.40 cm^3gm^{-1}.

Mo–1101 P: 11% MoO_3 on high activity alumina, as microspheroidal powder.

Mo–1201 T: 10% MoO_3 on high activity alumina, ⅛ inch tablets. The ABD is 0.85, the surface area 160 m^2gm^{-1}, and the pore volume 0.36 cm^3gm^{-1}.

Nalco Chemical Company
A 10% MoO_3 on alumina in powdered form suitable for use in a fluidized catalyst reactor.

III. Chromia–Alumina Reforming

As has been mentioned in Chapter 6, C_6 and higher paraffins are converted to aromatics by these catalysts. So are the cyclohexanes. In principle, this is a reforming process. However, none of the catalysts developed to this writing give as good a yield–antiknock quality combination as do the Pt–alumina reforming catalysts. It has not proved possible to operate these catalysts at elevated pressure and in the presence of H_2 recycle to obtain long operating periods, i.e., weeks or months, without regeneration. Specially prepared cogelled catalysts (*12*) permit reduced coke formation and thus increase product yield. For a while, this proved an attractive operation on high sulfur feeds because this catalyst was not only not poisoned by sulfur compounds, but sulfur was removed as H_2S. With improved hydrorefining processes and catalysts for removing sulfur and nitrogen compounds, this process has lost attractiveness compared to platinum reforming.

Typical operating conditions are (*19*): Temperature 500–550°C, pressure 6–15 atm, LHSV 0.3–0.7 $V_l V_c^{-1}$ hr^{-1}, and hydrogen circulation 3–10 moles per mole of hydrocarbon. The commercial operation was in moving bed units, using a cogelled 32% chromia–68% alumina catalyst in the form of beads. (The author is not aware of a current commercial supplier of this catalyst. Hughes *et al.* (*12*) have described the preparation of a similar catalyst.)

The retention time of the catalyst depended on the coke production and so varied. About 10 hours is a rough average.

A. Catalyst Poisons

The chromia–alumina catalysts are remarkably resistant to poisons. Much of the sulfur and nitrogen in compounds in the feed is removed as H_2S and NH_3 by these catalysts. Water however is a temporary poison, so the feed should be dry.

B. Regeneration

The chromia–alumina catalysts are less sensitive to temperature than the Pt–alumina catalysts. If provision is made to remove heat, they can be regenerated with air at temperatures in the 500–600°C range. Little or no CO is formed. While the catalyst is sensitive to moisture during the process, it is not necessary to dry the regeneration gas. During use and regeneration, the valence of the chromium changes. Some is present after regeneration as CrO_3 and can be extracted with water. While it is not likely that liquid water will come in contact with the regenerated catalyst, this fact should be kept in mind.

C. Suppliers of Proven Catalysts

See Chapter 6.

REFERENCES

1. Anonymous, *Hydrocarbon Process.* **47,** No. 9, 160, 161 (1968).
2. Anonymous, *Hydrocarbon Process.* **47,** No. 9, 156–158; 162 (1968).
3. Anonymous, *Hydrocarbon Process.* **47,** No. 9, 159 (1968).
4. Chen, N. Y., Maziuk, J., Schwartz, A. B., and Weisz, P. B., *Oil Gas J.* **66,** No. 47, 154 (1968).
5. Ciapetta, F. G., Dobres, R. M., and Baker, R. W., *Catalysis* **6,** 495 (1958).
6. Ewing, R. C., *Oil Gas J.* **67,** No. 14, 125 (1969).
7. Fowle, M. J., Bent, R. D., Ciapetta, F. G., Pitts, P. M., and Leum, L. N., *Advan. Chem. Ser.* **5,** 76 (1951).
8. Fowle, M. J., Bent, R. D., Milner, B. E., and Masologites, G. P., *Petrol. Refiner* **31,** No. 4, 156 (1952).
9. Haensel, V., *Am. Chem. Soc. Div. Petrol. Chem. Preprints* **6,** No. 3, A-93 (1961).
10. Haensel, V., and Berger, C. V., *Advan. Petrol. Chem. Refining* **1,** 386 (1958).
11. Haensel, V., and Sterba, M. J., *Advan. Chem. Ser.* **5,** 60 (1951).
12. Hughes, E. C., Stine, H. M., and Darling, S. M., *Ind. Eng. Chem.* **41,** 2184 (1949).
13. Jacobson, R. L., Kluksdahl, H. E., and Spurlock, B., U.S. Patent 3,434,960 (March 25, 1969).
14. Kluksdahl, H. E., U.S. Patent 3,415,737 (Dec. 10, 1968).
15. Kopf, F. W., Decker, W. H., Pfefferle, W. C., Dalson, M. H., and Nevison, J. A., *Oil Gas J.* **67,** No. 20, 141 (1969); *Hydrocarbon Process.* **48,** No. 5, 111 (1969).
16. McGrath, H. G., and Hill, L. R., *Advan. Chem. Ser.* **5,** 39 (1951).
17. Murphree, E. V., *Advan. Chem. Ser.* **5,** 58 (1951).
18. National Petroleum Refiners Association Forum, *Oil Gas J.* **66,** No. 16, 94 (1968).
19. Payne, J. W., Evans, L. P., Bergstrom, E. V., and Bowles, V. O., *Petrol. Refiner* **31,** No. 5, 117 (1952).
20. Steel, R. A., Bock, J. A., Hertwig, W. R., and Russum, L. W., *Petrol. Refiner* **33,** No. 5, 167 (1954).
21. Stormont, D. H., *Oil Gas J.* **67,** No. 17, 63 (1969).
22. Stormont, D. H., *Oil Gas J.* **67,** No. 15, 94 (1969).

CHAPTER 8

POLYMERIZATION

In this book the author is trying to provide the reader with proven catalysts which can be bought essentially ready to use. In polymerization, many catalysts are in commercial use that must be prepared just before use. The writer believes the reader would want to know about these and the related polymerizations so this chapter departs from the central theme by including these along with proven catalysts, even though some of the information may be less reliable. Free radical polymerizations will not be treated here.

I. Selective Polymerization of Propylene to Dimers

Depending on the catalyst and conditions, propylene can give many different dimers. A few catalyst systems have been developed which give predominently one dimer.

A. n-Hexenes

n-Hexenes are produced using HBr (introduced as isopropyl bromide) as a catalyst. At 450°C, 1 mole % of isopropyl bromide and 34 volumes of propylene per volume of reactor per hour produced a conversion of 22% per pass. The yield of *n*-hexenes was 80%, i.e., 45% 1-, 45% 2-, and 10% 3-hexene (*46*).

B. 2-Methylpentene-1

2-Methylpentene-1 is produced commercially as an intermediate in the production of isoprene [it is isomerized to 2-methylpentene-2, which is cracked to give isoprene plus methane (*3*)]. The catalyst is tripropylaluminum. Dried propylene–propane mixtures (60–80% C_3H_6) and the catalyst are charged to a two stage tubular reactor system under temperature and pressure conditions (not published) to give 70–85% C_3H_6 conversion. Other work (*70*) indicates that 2-methylpentene-1 is produced from propylene using tripropylaluminum as the catalyst at 140°C and elevated pressure, which doubtless varies with the partial pressure of propylene in the mixture.

C. 4-Methylpentene-1

4-Methylpentene-1 is produced from propylene, using finely divided potassium metal which reacts with the propylene to form the active catalyst. Reaction conditions are 150°C and 30–80 atm. The induction period of 1–12 hours may be the time required to form the catalytically active species. This induction period can be reduced by adding naphthalene to the system (*68*).

The catalyst is not soluble in the reaction system, so to obtain a large surface, the K is deposited as a thin film on a support (quartz sand, talc, α-alumina, or K_2CO_3). The reaction rate is 4–5 grams of C_3H_6 per gram of K per hour. The yield of 4-methylpentene-1 is about 80%. Maintaining an excess of C_3H_6 prevents side reactions and loss of catalyst activity.

A catalyst may be prepared as follows. Make a uniform paste of 15 gm of 1–6 μ talc and 125 ml of white mineral oil, for example, by ball milling. Dilute to 11% solids and add 0.5 ml of oleic acid. Under nitrogen, heat well above the melting point of K, e.g., to 130°C, add 15 gm of clean K, and stir violently until 99% of the particles are smaller than 10 μ (*68*). The author does not know of a supplier of the catalyst, although 4-methylpentene-1 is being made commercially (*10*).

D. 2,3-Dimethylbutenes

2,3-Dimethylbutenes and 2-methylpentenes are produced by closely related catalysts. The catalysts are homogeneous organometallic compounds which so far as the author knows are not available commercially. Since 2,3-dimethylbutenes have been made on a large scale using this type of catalyst, a description of a catalyst preparation is worthwhile. Suspend 13 gm of finely divided anhydrous $NiCl_2$ in 200 ml of chlorobenzene, and saturate with butadiene. Keep it saturated while the $NiCl_2$ is reduced with 2 moles of triethylaluminum. Remove the excess C_4H_6 by evacuation and, with good cooling, inject 5 moles of HCl gas into the solution. Then add slightly more than 1 mole of triphenylphosphine or triisopropylphosphine per gram atom of nickel. Finally, dilute to 0.1 molar in Ni with dry chlorobenzene. It is the phosphine added that determines the dimer formed. The triisopropylphosphine catalyst gives an 88% yield of C_6 olefins, of which 70% is 2,3-dimethylbutenes and 30% of 2-methylpentenes. The triphenylphosphine or the catalyst with no phosphine gives a 73% yield of 2-methylpentenes, 21% *n*-hexenes, and 5% 2,3-dimethylbutenes.

Reaction conditions are 30°–50°C, 15 atm pressure, with a feed rate of 2–6 kilograms of dry C_3H_6 per liter of catalyst solution per hour. Propylene conversion per pass is about 95%. As with many homogeneous reactions, the rate is fast, i.e., 13 kilograms of propylene reacted per gram of Ni per hour (*13*).

II. Oligomerization of Ethylene

One of Karl Ziegler's early observations was that an aluminum alkyl would react with an α-olefin to add the olefin to the alkyl group, thus giving a bigger alkyl group (*69*). Thus, $AlEt_3$ reacted with C_2H_4 to give tri-*n*-butylaluminum. On heating, 1-butene is split off, regenerating an alkyl aluminum hydride that is ready to repeat the cycle, i.e., a "catalyst."

In commercial operation, operating conditions are varied, depending on the molecular weights of the oligomer desired as well as the monomer. The temperature range is 100°–200°C and the pressure such as necessary to give the desired reaction rate, about 100 atm. Isopentane is a suitable solvent and heat transfer fluid when making C_6 and larger olefins.

In its simplest form the reaction can be represented as follows (*1*, *44*):

$$AlEt_3 + C_2H_4 \rightarrow AlEt_2Bu$$

The chain continues to grow randomly, giving a Poisson molecular weight distribution of aluminum alkyls. These are exchanged with ethylene in a second step to displace the oligomer olefins and regenerate the $AlEt_3$ for reuse. Nickel and cobalt are the common catalysts for this reaction. For example, adding a little nickel acetylacetonate to the mixed aluminum alkyls immediately gives a finely divided black suspension which is treated with C_2H_4 at 90–120°C and 10–100 atm. Special care must be used to remove the nickel catalyst from the product; otherwise it causes undesired reactions in the reuse of the aluminum alkyl catalyst.

The similarity of this process to that described above for making 2-methylpentene-1 from propylene should be noted.

1-Olefins of higher molecular weight respond to the same catalysts but are usually more difficult to oligomerize. Thus, when copolymers of a higher olefin and ethylene are wanted, the ratio of ethylene to the other monomer is usually smaller than the ratio wanted in the oligomer, because of the tendency of the ethylene to polymerize faster.

A. Catalyst Suppliers

Both Ethyl Corporation and Texas Alkyls, Inc. supply aluminum alkyls as well as alkyl aluminum chlorides.

III. Polybutenes

In a much older process, olefins, especially butenes, may be polymerized to olefinic liquids with molecular weights in the range of 300–2500 ("polybutenes"). The catalyst is a Friedel-Crafts catalyst; anhydrous aluminum chloride + HCl or an alkyl chloride is used. The temperature is in the 50°–100°C range and the technique is the same as that ordinarily used with $AlCl_3$. The reaction probably takes place in the catalyst–"sludge" layer (*5, 60*).

IV. Olefinic Gases to Gasolines

The catalytic polymerization of propylene and butylenes to liquid, gasoline boiling range polymers is one of the oldest catalytic processes used in petroleum refining. Phosphoric acid is the active catalyst. The most common catalyst is the "solid phosphoric acid" catalyst based on V. N. Ipatieff's work and is about 60% P_2O_5–40% SiO_2 (kieselguhr). A film of phosphoric acid on crushed silica has also been used. This permits replacing the phosphoric acid as it becomes contaminated. Operating conditions are: 175°–235°C (usually 200°–220°C) and 25–80 atm pressure; space velocity is 2.5 liters per kilogram of catalyst per hour. The primary effect of space velocity is to alter olefin conversion per pass, i.e., an increase in space velocity causes a decline in monomer disappearance rate and vice versa. Space velocity has very little effect on product distribution, e.g., an increase in space velocity has very little effect on dimer:trimer ratio in propylene polymerization.

The energy of activation of the reaction is small, i.e., about 5 kcal $mole^{-1}$. This, together with the fact that the surface area of the solid catalyst is about 2–4 m^2gm^{-1}, suggests that the reaction may be diffusion controlled as ordinarily operated commercially.

The reaction is exothermic (25–30 kcal per gram mole of dimer); thus, heat removal can more nearly determine feed rate than reaction rate. The reaction is conducted in chambers with cold recycle to control temperature rise, or in tubes (2–4 inches in diameter) with a heat transfer fluid (usually water boiling under pressure) on the outside.

The catalysts can be deactivated by too much water or too little water in the feed. Too little water dehydrates the phosphoric acid to polyphosphoric acids which are less active. Too much water softens the solid phosphoric acid catalyst causing it to compact and increase pressure drop across the reactor. Water in the feed should be kept so the catalyst contains between 15 and 18% "free" P_2O_5 (P_2O_5 that can be leached from the catalyst by cold water in the laboratory). The amount of water in the feed to produce this will vary with the operating temperature. 350–400 ppm H_2O in the feed is the range for 200°–220°C. If polymerization is conducted at too low a pressure, the polymer will be almost water white with a slight fluorescence but the catalyst will deactivate rapidly due to the accumulation of "polymer" on the surface. Higher pressures give a condensed phase ("pseudo" liquid) which washes heavy polymer off the catalyst, produces a yellow product, and retards deactivation. Polymer recycle (used to make tetramer and pentamer) also can help keep the catalyst clean unless conditions are too severe.

As the temperature is raised above 220°C, more "conjunct" polymerization occurs, i.e., a hydrogen transfer reaction in which a paraffin molecule is formed along with a diolefin which polymerizes to deactivate the catalyst or even forms some aromatic hydrocarbons.

Relative rates of polymerization are C_3H_6 = 1; 1-C_4H_8 = 2; *i*-C_4H_8 = 10. In a mixture containing only C_4 hydrocarbons, this provides a means of selectively polymerizing *i*-C_4H_8 and leaving the other two butenes. To do this selective polymerization, the temperature is lowered about 25°–30°C. In the presence of polymerizing *i*-C_4H_8, some *n*-butenes polymerize to form a codimer.

The polymerization of butenes to make gasoline has been largely replaced by the alkylation of isobutane with the butenes. The polymerization of propylene is declining too, as more of it is being used as a chemical raw material.

While *i*-C_4H_8 gives primarily diisobutylene, a simple mixture, the other polymers made with the phosphoric acid catalyst are complex mixtures. C_3H_6 gives mostly trimers and tetramers; the dimer yield rarely exceeds 10%. Many research laboratories have examined the propylene polymers seeking simple, useful petrochemical intermediates without success. The complexity of the product is due in part to some of the practical considerations imposed on the process. The process is operated to get the highest possible conversion of C_3H_6 to gasoline. This means that primary reaction products are held in contact with the catalyst for a long time to give the residual C_3H_6 a chance to react. Thus, the primary products have ample opportunity to undergo secondary reaction, and they do. This is a common occurrance in catalytic reactions, a large conversion per pass often leads to secondary reactions (*4*, *35*, *45*).

A. Poisons

As would be expected from the acidic nature of the catalyst, organic bases, especially, and NH_3 in the C_3H_6 feed are poisons. In addition, oxygen (>20 ppm) and butadiene in the feed cause catalyst deposit and catalyst deactivation.

B. Regeneration

The solid phosphoric acid catalysts are not easily regenerable so spent catalyst is replaced with fresh. This is economical since polymer production is 400–1600 liters per kilogram of catalyst.

B. Suppliers of Proven Catalysts

Catalysts and Chemicals, Inc.
C84: as 3/16 to 3/8 inch spheres.

Universal Oil Products Co.
No. 1: as 1/4 inch extrudates and No. 2 and No. 4: as 3/16 inch extrudates.

V. Dimerization of Higher Olefins

The dimerization of higher olefins using aluminum alkyl catalysts is general and can be illustrated by the dimerization of 4-methylpentene-1, which has been operated on a pilot plant scale (*31*). The catalyst was triisohexylaluminum (made by exchanging between $Al(i\text{-}Bu)_3$ and 4-methylpentene-1 at 120°C). The reaction does not take place in the vapor phase, so sufficient pressure must be used to maintain a liquid phase. The reaction was conducted by passing a solution of the catalyst in the olefin through an empty heated tube. As a result of a process variable study, the following optimum conditions were found. Temperature 218°C, olefin : catalyst molar ratio 13 : 1, LHSV 2.78 $V_l V_0^{-1}$ hr^{-1}, dimer in the product 30.5%, and space time yield 450 grams of product per liter of reactor volume per hour. Above about 30 atm, pressure was not an important variable.

The main reaction product was 2-isobutyl-6-methylheptene-1, 65–87%. 2-Isobutyl-6-methylheptene-2 was 11–20% and 2,9-dimethyldecene-3, -4, and -5 were 11–15%.

VI. Olefin Disproportionation (Olefin Dismutation)

This interesting process, while not polymerization, may be related to both polymerization and depolymerization. In the process, *n*-olefins are converted into olefins containing one more and one less carbon atoms than the original olefin. In its simplest form it can be represented (*11*):

$$2\ C_3H_6 \rightleftharpoons C_2H_4 + C_4H_8$$

and at least one commercial plant uses it for this conversion (*42, 43*).

The reaction has been demonstrated for propylene, butenes, pentenes, hexenes, heptenes, and octenes. Isobutylene gives ethylene and 2,3-dimethylbutenes. The reaction has several interesting features which are important to selecting the right catalyst for the desired result.

(1) The reaction is reversible (*15*).

(2) The catalyst may have activity for shifting double bonds. Thus, 1-pentene gives ethylene plus 4-octene. However, if the catalyst also shifts double bonds, the product may be (*a*) a mixture of octenes, because the double bond has been moved, or it may be (*b*) 3-hexene plus 2-butene, because the 1-pentene was isomerized to 2-pentene before reacting. Adding a little base, e.g., $NaHCO_3$ to the catalyst inhibits the isomerization so that a more selective reaction occurs (*15*). A Na_2O content of 0.2% in the catalyst is adequate.

(3) The double bond isomerization may be desirable and need enhancing. For example, 2-butene reacts with itself to regenerate 2-butene, i.e., no change, and the catalyst seems to have little or no activity. By adding an isomerization catalyst, e.g., MgO, to one of the catalysts (WO_3 on silica gel), the apparent activity was increased because the 2-butene was isomerized to 1-butene which then reacted to give ethylene + 3-hexene.

(4) The fact that the reaction is reversible means that the two reacting olefin molecules need not be identical so that some interesting products can be obtained. This could result in a complex mixture; however, ethylene cannot disproportionate, so it is a potentially useful reactant (*15*). For example, ethylene + 2,3-dimethylbutene gave isobutylene (*20*).

Cycloolefins also react, e.g., with ethylene and give α,Ω-diolefins, e.g., 1,7-octadiene was obtained from cyclohexene plus ethylene (*12*).

A number of catalysts has been described for the process (reaction). The three most useful seem to be commercial cobalt oxide–molybdenum oxide–alumina (the so called "cobalt molybdate" catalysts) (*11*), 7% WO_3 on wide pore silica gel (*32*), and Re_2O_7 on alumina (*12*). WO_3 on silica gel is active in the 400°C range, "cobalt molybdate" in the 150°–200°C range, and Re_2O_7 on alumina at 30°–50°C. Of these catalysts, so far as the writer knows, the "cobalt molybdate" catalyst is the only one available commercially, so catalyst treatment, operating conditions for propylene to ethylene + butenes, and poisons will be given for it.

The catalyst (with or without alkali added) is heated at 550°–580°C for 1 hour in *dry* air and 1.5 hours in dry nitrogen, preferably in the reactor in which it is to be used. If this latter is not possible, the catalyst is transferred hot to its container and all catalyst transfers made under dry box conditions (*15*). The operating temperature is 150°–200°C, atmospheric pressure, and GHSV 500–2000 $V_gV_c^{-1}$ hr^{-1} in the laboratory.

To provide economic reactor space, commercial operation is at pressures up to 30 atm, temperatures of 300°–475°C and WHSV 10–100 $W_rW_c^{-1}$ hr^{-1} to give propylene conversions of about 43% (about equilibrium) per pass. The mole ratio of C_2H_4:C_4H_8 is around 1.1. Some higher olefins, e.g., hexenes, are produced (*11*, *42*, *43*).

Although not used commercially as of this writing, diolefins and acetylenes also react. For example, with the 7% WO_3 on silica gel catalyst (0.19% Na_2O), butadiene gave ethylene and cyclohexadiene (instead of the expected hexatriene) at 538°C and a WHSV of 26 $W_rW_c^{-1}$ hr^{-1} (*12*, *33*).

A. Poisons

Water is a temporary poison, so the feed must be dried. Accidental moisture poisoning can be corrected by continuing dry feed. Sulfur compounds behave similarly, so the feed should be desulfurized. Acetylenes and diolefins slowly deactivate the catalyst for olefin conversion (*42*, *43*).

B. Suppliers of Proven Catalysts

American Cyanamid Co.

Aero HDS–2: 3% CoO, 15% MoO_3, 0.025% Na_2O on alumina as 1/16 and 1/8 inch diameter particles. The apparent bulk density (ABD) is 0.5, the surface area 275 m^{2}gm^{-1}, and the pore volume 0.65 cm^{3}gm^{-1}.

Aero HDS–6: 4% CoO, 10% MoO_3, 0.025% Na_2O on alumina. The ABD is 0.5, the surface area and pore volume the same as HDS–2; 1/16 and 1/8 inch particles.

Catalysts and Chemicals, Inc.

C20–6, a CoO, MoO_3 on alumina catalyst; 1/16 and 1/8 inch particles.

Catalysts Division of Chemetron Corporation

Girdler G–35: tabletted 3.5% CoO, 10% MoO_3 on alumina.

Girdler G–51: extruded 3.5% CoO, 10% MoO_3 on alumina.

Girdler G–85: spherical 2.0% CoO, 10% MoO_3 on alumina.

Filtrol Corporation

Grade 475–8: 2.8% CoO, 15% MoO_3 on alumina, as 1/8 inch extrudates. The ABD is 0.75.

Harshaw Chemical Company

CoMo–0401 T: a silica promoted catalyst containing 3% CoO and 9% MoO_3 on alumina, as ⅛ inch tablets. The ABD is 1.0, the surface area 160 m^2gm^{-1}, and the pore volume 0.40 cm^3gm^{-1}.

CoMo–0402 T: a silica promoted catalyst containing 3% CoO and 15% MoO_3 on alumina, as ⅛ inch tablets. The ABD is 1.0, the surface area 200 m^2gm^{-1}, and the pore volume 0.40 cm^3gm^{-1}.

CoMo–0601 T: a nonsilicated catalyst containing 3% CoO and 9% MoO_3 on alumina, as ⅛ inch tablets. The ABD is 1.1, the surface area 160 m^2gm^{-1}, and the pore volume 0.40 cm^3gm^{-1}.

CoMo–0603 T: a nonsilicated catalyst containing 3% CoO and 12% MoO_3 on alumina, as ⅛ inch tablets. The ABD is 1, the surface area 166 m^2gm^{-1}, and the pore volume 0.40 cm^3gm^{-1}.

HT–200: 3% CoO plus 12% MoO_3 on alumina as 1/32, 1/16, 1/12, and ⅛ inch extrudates. The surface area is 180 m^2gm^{-1} and the pore volume 0.42 cm^3gm^{-1}.

Houdry Process and Chemical Company

⅛ and 5/32 inch extrudates of 3.2% CoO, 14.6% MoO_3, and 0.03% Na_2O on alumina. The ABD is 0.75 and the surface area 330 m^2gm^{-1}.

Nalco Chemical Company supplies the following as 1/16 and ⅛ inch extrudates:

Nalcomo 471: 3.5% CoO, 12.5% MoO_3, 0.05% Na_2O. The ABD is 0.64, the surface area 270 m^2gm^{-1}, and the pore volume 0.48 cm^3gm^{-1}.

Nalcomo 810: 3.5% CoO, 12.5% MoO_3, 0.05% Na_2O. The ABD is 0.68, the surface area 153 m^2gm^{-1}, and the pore volume 0.32 cm^3gm^{-1}.

Universal Oil Products Company

Hydrobon S–3: "cobalt molybdenum" as ⅛ inch pellets; the ABD is 1.

VII. α-Olefinic Hydrocarbons to Solids

A. Ethylene to Polyethylene (Ziegler)

Dr. Karl Ziegler of the Max Planck Institut began a revolution in the plastics industry with his discovery of catalyst systems that would polymerize ethylene to plastics at low pressures (*71*). Before that, polyethylene had been made at high pressures (1000 atm) by a thermal free radical process.

Each commercial user of the Ziegler process has his own favorite catalyst system. The following is given as an illustration.

A catalyst is prepared from an aluminum alkyl, e.g., $AlEt_3$ or $Al(iBu)_3$, and titanium tetrachloride, with an excess of the alkyl. The preparation is carried out in a dry hydrocarbon solvent, usually hexane. The precipitate + the aluminum alkyl still in solution is the catalyst.

With stirring (in a vessel equipped to remove the heat of reaction), ethylene is added to a pressure of about 10 atm and the temperature controlled to 50°–75°C. The polymer forms as a white powder suspended in the "solvent."

At the end of the reaction, methanol is added to kill the catalyst, the catalyst residue extracted from the polymer by HCl or methanol + HCl, and it is water washed. It is then dried, antioxidants are added, and it is extruded into pellets.

Catalyst consumption is less than 0.1% of $TiCl_4$ and 1% aluminum alkyl based on polymer.

A small amount of soluble polymer left in the solvent is removed by distillation before the solvent is dried and reused (*51*).

1. *Poisons*

Because the aluminum alkyls are so reactive, ethylene and solvent must be essentially free of H_2O, O_2, CO, CO_2, SO_2, COS, etc.

2. *Catalyst Suppliers*

The Anderson Chemical Division of Stauffer Chemical Co. and the Ethyl Corp. supply aluminum alkyls and alkyl aluminum chlorides.

B. Ethylene to Polyethylene (Amoco Chemicals Company)

The purified monomer in a solvent for polyethylene (cyclohexane, decalin, benzene, xylenes, or "odorless spirits") reacts at 240°–300°C and 35–100 atm pressure in the presence of a mixed oxide plus cocatalyst to give solid polymers. At the end of the polymerization, excess monomer is removed, the polymer solution filtered to remove catalyst, the catalyst washed with hot solvent to recover more polymer, and the solid polymer recovered from the solvent (*7*, *25*, *27*). Specific details of the commercial operation are not available.

For the researcher, the following may be helpful (*23*, *24*, *39*, *57*, *58*). Carefully dried petroleum, paraffinic naphtha, can be used as a solvent. The ethylene must be purified to less than 5 ppm of each H_2S, CO_2, CO, O_2, and the dew point must be of −45°C or less. The catalyst, 7–25% molybdic oxide on 20/200 mesh γ-alumina, is calcined at 500°–600°C, and reduced by hydrogen at 400°–600°C. It is used without exposing it to the air. Powerful reducing agents are cocatalysts, e.g., $LiBH_4$, $AlEt_3$, Et_2AlCl, or metallic Na. Metallic Na is a convenient cocatalyst, since the operating temperature is above its melting point and it can be introduced as a solid or liquid. The ratio of molybdenum oxide–alumina to Na is about 20:1–2:1. At the operating temperature, a pressure of about 100 atm is used to keep the solvent in the liquid phase and to maintain an adequate partial pressure of ethylene. On a small scale, 30–50 gm of polymer is formed per gm of molybdenum oxide–alumina, but the writer understands continuous operation gives yields approaching ten times this. The polymerization is carried out with agitation and continued until the polymer concentration is about 10–25% in the solvent. After removing excess ethylene, the hot solution is filtered to remove catalyst and the catalyst washed with additional hot solvent to recover polymer from the catalyst. The catalyst may be reused as is or reacted and extracted with alcohol to remove Na, oxidized with air to burn out polymer residues, and processed like fresh molybdenum oxide–alumina.

1. *Catalyst Suppliers*

Harshaw Chemical Company

Mo–0502 T: as ⅛ inch tablets of 10% MoO_3 on activated alumina. The ABD is 1, the surface area 64 m^2gm^{-1}, and the pore volume 0.40 cm^3gm^{-1}.

Mo–01201 T: as ⅛ inch tablets of 10% MoO_3 on high activity alumina. The ABD is 0.8, the surface area 160 m^2gm^{-1}, and the pore volume 0.36 cm^3gm^{-1}.

Nalco Chemical Company furnishes a molybdena on alumina catalyst.

C. Ethylene to Polyethylene (Phillips Petroleum Co.)

In many respects, the Phillips Petroleum Co. and the Amoco Chemical Co. polyethylene processes resemble each other, except for differences in the catalyst. The Phillips process uses a catalyst of 2–3% CrO_3 on a silica–alumina cracking catalyst. The silica–alumina (90% SiO_2–10% Al_2O_3) is steam aged at 480°–650°C to destroy small pores, increase the pore size, and decrease surface area. It is then impregnated with CrO_3 and activated by heating in air at 500°C. (The catalyst is highly sensitive to moisture so that the dew point of any air in contact with it should be −40°C or lower.) During polymerization, no O_2, H_2O, or CO_2 should be in the C_2H_4 or solvent (*8*, *17–19*, *25*).

Cyclohexane is the usual solvent. 0.3–0.7% of catalyst (based on weight of solvent) is slurried into the solvent and heated to 100 –150°C in the reactor. C_2H_4 is absorbed in the solvent at 35 atm and charged to the reactor (35 atm). The rate of addition of the solution of ethylene is controlled to control the temperature in the exothermic reaction. The polyethylene formed dissolves in the solvent to produce a viscous solution.

At the end of the reaction, the pressure is reduced to 7 atm and the C_2H_4 released is recycled to the C_2H_4 absorber for reuse. The hot polymer solution is centrifuged, then filtered to remove catalyst particles. The polymer separates as a solid when the clean solution cools.

The catalyst is regenerated by washing with solvent at 150°–200°C and then burned with air at 480°–600°C.

1. *Catalyst Supplier*

Davison Chemical Division of W. R. Grace & Co. manufacturers the catalyst, and they have an agreement with Phillips Petroleum Co. not to sell it to any one except licensees of the process. The catalyst is not too difficult to prepare and the author knows of researchers who have done successful experimentation on "homemade" catalyst.

D. Higher α-Olefinic Hydrocarbons to Plastics

Propylene and the larger α-olefins are more difficult than ethylene to polymerize. Professor G. Natta found a series of catalysts which would not only polymerize propylene and higher olefins but would produce highly ordered, crystalline polymers. For example, "isotactic" polypropylene has the carbon atoms in each succeeding C_3H_6 unit ordered in exactly the same way as the preceding unit.

Figure 2 represents the ordered structure of isotactic polypropylene.

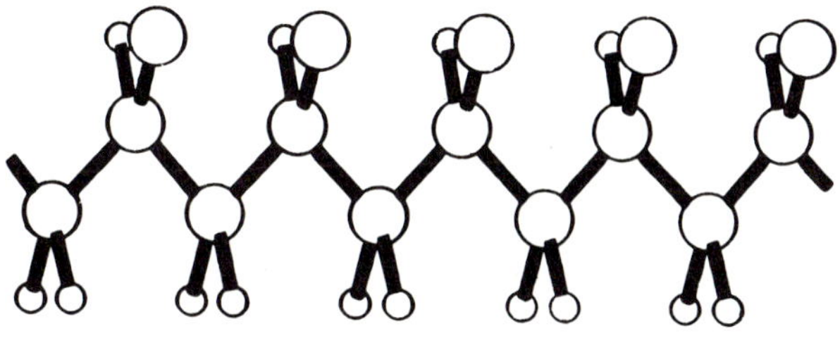

Fig. 2. The ordered structure of isotactic polypropylene.

(Professor Natta once told the writer that his wife had suggested the term "isotactic" to characterize this structure. He was describing to her the structure he was finding by X-ray diffraction when she said it reminded her of the ordered lines and columns of marching soldiers, hence "isotactic").

In summarizing his work (*50*, *51*) Professor Natta showed the necessity of having a crystalline or microcrystalline catalyst to produce stereospecific results. Crystalline α-$TiCl_3$ (often ball milled in the solvent to produce small crystals) + an aluminum alkyl or a dialkylaluminum chloride is probably the simplest catalyst system. Again, the aluminum alkyl must be in excess so that some will be present in the solvent. In commercial operation, individual manufacturers have their own preferred proprietary catalyst preparations and most of them have a third component added.

The operating conditions used for Ziegler polyethylene will work for polypropylene although somewhat higher temperatures and pressures are preferred commercially to increase the reaction rate, e.g., 75°–125°C, and 10 atm pressure of propylene. Hydrogen pressure is often used to control the molecular weight of the product, i.e., increasing hydrogen pressure decreases molecular weight. Thus, total pressures up to 20 atm are not unusual. Both batch and continuously stirred reactors are used, with several reactors in series being used with the continuous reactor systems.

Stopping the reaction, extracting catalyst, etc., are processes similar to that used for polyethylene. Depending on the exact catalyst, from 1 to about 10% of the converted propylene goes to an amorphous, atactic polymer which is soluble in the solvent and is removed by distillation before reusing the solvent.

As of this writing the Ziegler–Natta catalyst system is the one being used widely in making polypropylene and for polymerizing higher olefinic hydrocarbons. The writer is not able to give more specific details on propylene polymerization.

1. *Poisons*

The same as those for polyethylene.

2. *Suppliers of Proven Catalysts*

A few manufacturers make their own aluminumalkyls from metallic aluminum, the olefin, and hydrogen. A little alkyl from a previous batch is used to start the reaction.

Ethyl Corporation supplies aluminum alkyls and alkylaluminum chlorides.

Anderson Chemical Division of Stauffer Chemical Co. supplies the aluminum alkyls as well as the following grades of titanium trichloride:

Grade A ($TiCl_3 \cdot AlCl_3$): Aluminum reduced.

Grade AA ($TiCl_3 \cdot AlCl_3$): Aluminum reduced and activated.

Grade H ($TiCl_3$): Hydrogen reduced.

Grade HA ($TiCl_3$): Hydrogen reduced and activated.

E. Ethylene–Propylene Copolymers (Plastics)

Ethylene polymerizes so much more easily than propylene or higher olefins that it is difficult to get true copolymers. By resorting to "tricks," it is possible to get useful copolymer products. One trick uses extremely low concentrations of ethylene compared to propylene. Another forms "block" polymers by adding alternating slugs of monomers so that the growing polymer chain contains segiments of one type of polymer and then the other. Such polymers have desirable properties not possessed by either pure polymer.

Operating conditions are those used for polymerizing propylene. As of this writing, the Ziegler–Natta type catalyst systems are the most commonly used ones for copolymerizing ethylene and propylene. If the Amoco Chemicals Co. catalyst version using Na is employed, it should be noted that xylene is not a suitable solvent, for some of the propylene is consumed in alkylating the side chains of xylene.

1. *Poisons*

The same as those for polyethylene catalysts.

2. *Suppliers*

The same as for polyethylene and polypropylene.

F. Ethylene–Propylene Copolymers (Rubbers)

Replacing titanium with a different transition element, vanadium, gives amorphous elastomers instead of crystalline solid plastics (*52*). Vanadyl chloride or alkyl vanadates are the sources of vanadium and they produce homogeneous catalysts. This may be the reason the reaction rate is so much faster than in the titanium polypropylene system. Hexane, heptane, or highly chlorinated aliphatics, such as tetrachloroethylene, are solvents. The concentration of V compound in the solvent is 0.1–0.5 millimoles liter^{-1}. The Al:V ratio is 10–15:1 and $AlEt_3$, Et_2AlCl, and $EtAlCl_2$ are all useful, with some preference shown for mixtures of Et_2AlCl and $EtAlCl_2$, e.g., 50:50. Any excess $AlCl_3$ left over from preparing the alkyl aluminum chloride is damaging. The reaction goes at room temperature with the exothermic heat being controlled by cooling and olefin rate addition. With adequate cooling, a batch can be completed in 20–30 minutes after the polymerization starts. The run is terminated when the solution of polymer gets too viscous for effective stirring and cooling.

Ethylene polymerizes faster than propylene so that the rate slows as the propylene content of the feed increases. Around 50–60% propylene in the feed is a suitable composition (*38*).

Ethylene–propylene rubber is essentially paraffinic and so is not vulcanizable with the usual agents. By incorporating suitable dienes in the polymer, the usual vulcanizing techniques can be used. From 0.3 to 0.6 moles of unsaturation per kilogram of polymer is adequate (*26*). Dienes with both double bonds terminal are not suitable for both bonds enter into the polymer and so do not provide unsaturation. With one terminal double bond and one internal, the terminal double bond reacts to incorporate the molecule in the polymer chain, while the nonreacting internal double bond provides the unsaturation for vulcanization. The double bonds in the diene are preferably far apart, for, as they come closer together, they inhibit the polymerization. Norbornadienes, at 0.01–0.02 moles liter^{-1} of solvent, copolymerize at a rate that will provide a polymer with adequate unsaturation, while three to six times the concentration is required for the linear dienes. The three widely used dienes are ethylidene norbornene, 1,4-hexadiene, and dicyclopentadiene (*26*).

The following detailed procedure illustrates the technique (*26*). The dried and deaerated solvent is charged to the stirred reactor, followed by the diene. The addition of ethylene and propylene in the desired ratio is started to saturate the solvent. Next, the aluminum alkyl is added, followed in a few minutes by the vanadyl chloride. Polymerization begins immediately, and the temperature rises. With hexane or benzene as a solvent, refluxing solvent can be part of the cooling technique. Polymerization is continued until the mass becomes too viscous to stir effectively. The catalyst is killed with methanol and the polymer isolated from the solvent.

To produce useful rubbers, the upper limit of ethylene is about 80 mole % and the lower limit about 33% (*26*).

1. *Poisons*

The same as for polyethylene.

2. *Catalyst Suppliers*

Anderson Chemical Division of Stauffer Chemical Co. is a supplier of vanadyl chloride.

VIII. Isobutylene to Rubber

Isobutylene and 2–3% isoprene may be copolymerized at −75° to −100°C by a small amount of a Friedel–Crafts catalyst. The product, Butyl rubber, contains enough unsaturation from the isoprene to permit vulcanization. The isoprene is not essential to polymerization, so a rubber can be made from isobutylene alone, but it is not vulcanizable with usual agents.

Isobutylene plus isoprene dissolved in liquid ethane is polymerized at boiling ethane temperature by a small amount of $AlCl_3$ dissolved in C_2H_5Cl, or BF_3. The reaction is extremely fast, reminding one of ionic reactions in inorganic chemistry. The reaction is also exothermic, so much of the technology is involved with absorbing large sudden heats of reaction at low temperatures (*65*). Boiling off part of the liquid ethane is one method of absorbing this heat.

A. Poisons

Water in the solvent or feed, as well as other compounds that react with $AlCl_3$ or BF_3, should be avoided.

IX. Butadiene to Liquids

1,5-Cyclooctadiene and 1,5,9-cyclododecatriene are butadiene dimers and trimers, respectively. Although they are commercial products made by polymerizing butadiene (*9*), the author does not know the specific catalyst(s) used. The following summary may be interesting, since the catalysts are homogeneous nickel catalysts with the nickel probably in the zero ("bare nickel") state.

Wilke (*67*), during the study of the diethylaluminum chloride–titanium tetrachloride system for polymerizing butadiene to 1,4-*trans*-polybutadiene, found that increasing the Al:Ti ratio from 1:1 to 4.5:1 changed the reaction so that *trans*,*trans*,*cis*-1,5,9-cyclododecatriene was formed.

Bisacrylonitrile bistriphenylphosphine nickel (*63*) or tetraphenylisonitrile nickel triphenyl phosphite at 80°C and 10 atm pressure in a cyclooctadiene (COD) or benzene solvent polymerize butadiene to cyclododecatriene (CDT) (*55*).

Nickel catalysts with other ligands give COD. Bisacrylonitrile nickel (*62*) gives this dimer (*55*) at 50°–100°C and around 10 atm pressure. Benzene or COD are solvents. It is also produced using bistriphenyl phosphite dicarbonyl nickel at 125°–200°C and 30–60 atm pressure with a catalyst concentration of 1–3% (*14*, *61*). [The catalyst was prepared from $Ni(CO)_4$ and triphenyl phosphite. It is possible to prepare such catalysts from the less toxic biscyclopentadienyl nickel(II) or biscyclopentadienyl nickel(III) salts and triaryl or trialkyl phosphites (*54*).]

Linear polymers, e.g., 1,3,7-octatriene, are formed if the reaction is carried out with the Ni^0 catalysts in the presence of an alcohol or phenol. Somewhat analogous Co catalysts produce 3-methyl-1,4,6-heptatriene (*22*, *48*). So far as the writer knows, these are not now commercial products.

X. Dienes to Rubbers and Plastics

It is possible to polymerize butadiene "head-to-tail" (1,4 polymerization) so that the residual $—CH_2CH{=}CHCH_2—$ units are cis or trans. When they are cis, the product is a rubber and, when trans, a plastic. Butadiene may also be 1,2 polymerized, i.e., a vinyl polymerization. It is the 1,4-cis polymer that is of primary commercial importance.

Isoprene can polymerize in analogous ways and, in addition, can form 3,4 polymers. It is the 1,4-cis polymer ("synthetic natural rubber") that is of greatest utility, followed by the 1,4-trans (synthetic balata).

Many factors come into play in determining the structure of the polymer, such as ratio of components in the catalyst, temperature, and solvent. Furthermore, the same catalyst system may or may not give the same configuration with both butadiene and isoprene. Table II summarizes some

TABLE II

EFFECT OF CATALYST VARIATIONS ON THE STRUCTURE OF BUTADIENE AND ISOPRENE POLYMERS[a]

	Polybutadiene			Polyisoprene		
Catalyst	*t*-1,4-	*c*-1,4-	1,2-	*t*-1,4-	*c*-1,4-	3,4-
$AlR_3 + TiCl_4$	x[b]				x	
$AlR_3 + TiCl_3$	x			x		
$AlR_3 + TiI_4$		x		—	—	—
$AlR_3 + VCl_4(VCl_3)$	x			x		
$AlR_3 + Ti(OR)_4$			x			x
$AlR_3 + Ti(acac)_3$[c]			x			x
$LiAlH_4 + TiCl_4$			x	na[c]	na[c]	na[c]
$LiAlH_4 + TiI_4$	x			x		
$CdR_2 + TiCl_4$	x				x	

[a] See reference (*28*).
[b] x = major product.
[c] acac = acetylacetonate; na = not available.

of these effects. It has been noted that butadiene itself contains some 96% of the trans isomer at room temperature, while isoprene contains some 85% of the cis isomer at 50°C, and that this difference might be related to some of the seeming anomalies in the effect of different catalysts (*28*). Catalysts that give a configuration different from that of the monomer would not only have to polymerize but also isomerize.

Lehr (*41*) has published a review that contains more detail than can be included here.

A. Cis-1,4-polybutadiene

This polymer may be made using the following Co + Et_2AlCl catalyst system. All operations must be conducted under dry box conditions, with dry reagents and solvents and oven dried containers. Prepare a 0.5 *M* suspension of $CoCl_2$ in dried benzene and add 2 moles of pyridine (dried) per mole of $CoCl_2$. Stir for 3 hours, filter through sintered glass, and use the filtrate. The solubility of $CoCl_2 \cdot 2py$ is about 6.8×10^{-4} moles per liter of benzene. The Et_2AlCl is 1 *M* in benzene. For controlled water addition, prepare a saturated solution of water in benzene.

At 0°C, prepare a 10% solution of dry C_4H_6 in benzene and add 20 mmoles of Et_2AlCl per 100 gm of C_4H_6. Next, add 0.02 mmoles of $CoCl_2 \cdot 2py$ and then 5 mole % H_2O based on the Et_2AlCl. Polymerization begins with the addition of the water. Polymerize 6 hours at 5°–6°C and then add isopropanol to kill the reaction and precipitate the rubber. Leach 24 hours with methanol, and dry. For significant tests on the rubber, an antioxidant, e.g., phenyl-β-naphthylamine should be present in the isopropanol. Yield 95+% or 95+% *cis*-1,4-polybutadiene (*30*).

Polymerization to *cis*-1,4-polybutadiene with aluminum alkyls and a soluble cobalt catalyst is a commercial process (*6, 21*). Cobalt chelates provide another way of introducing soluble cobalt to make homogeneous catalysts (*36*). The small amount of cobalt needed in these polymerizations is noteworthy.

$AlEt_3$ or $Al(iBu)_3$ + TiI_4 forms a heterogeneous catalyst which produces *cis*-1,4-polybutadiene (*59*), as does aluminum alkyls plus β-titanium trichloride (brown) (*49*). In contrast, aluminum alkyls plus α-titanium trichloride (violet) produce *trans*-1,4-polybutadiene.

It is intriguing that reduced Ni on kieselguhr (a hydrogenation catalyst) or Ni on acidic supports (e.g. silica–alumina) produces *cis*-1,4-polybutadiene at 40°–50°C. The acidic support seems necessary in that Raney nickel did not work (*66*).

B. Trans-1,4-polybutadiene

The aluminum alkyl–titanium tetrachloride catalyst system used for making polyethylene will polymerize butadiene. Dry benzene is the preferred "solvent." To make *trans*-1,4-polybutadiene, the Al:Ti ratio is critical. For example, with $Al(iBu)_3$ and $TiCl_4$ the Al:Ti ratio must be 1.25 or less, using benzene for the solvent at 3°C, and must be less than 1.0 using heptane as the solvent at 25°C (*28*). Et_2AlCl + $TiCl_4$ with Al:Ti 1:1 or less also gives *trans*-1,4-polybutadiene (*67*).

Et_3Al + $VOCl_3$ or VCl_3 gives the trans product at 15°C, as does Et_3Al + α-titanium trichloride (violet) (*49*).

C. 1,2-Polybutadiene

The following catalyst system is interesting for two reasons: (*a*) It uses no aluminum alkyl and (*b*) it produces almost pure 1,2-polybutadiene. The catalyst is made from $Co_2(CO)_8$ and $MoCl_5$; Co:Mo is 1:1. To a benzene solution of $Co_2(CO)_8$ (0.1 *M*), add with stirring the needed $MoCl_5$ as a suspension in heptane (0.1 mole liter^{-1}) at room temperature. CO is evolved and a black precipitate forms immediately which is the catalyst. It is aged at 50°C for 30 minutes and used without separating. Oxygen must be excluded, but a small amount of H_2O is essential for activity (1.7–3.5 mole % of the $MoCl_5$). The polymerization is carried out at 50°C, with the concentration of C_4H_6 in the benzene solvent being 1.7 *M*, plus 2.5 mmoles of Co (or Mo) per mole of C_4H_6. The reaction time is 18 hours, after which the reaction is stopped with methanol and the product separated from the reaction mixture (*56*).

The 1,2 polymer is formed from Et_3Al + alkyl titanates or titanium(III) acetylacetonate. It is also formed from $LiAlH_4$ and $TiCl_4$ (*28*).

D. Cis-1,4-polyisoprene

Although the Ziegler AlR_3–$TiCl_4$ catalyst gives the trans polymer with butadiene, it gives the cis polymer with isoprene. To a 75:25 solution of dry petroleum ether–dry isoprene at 7°C, add the catalyst from $AlEt_3$, or $Al(iBu)_3$ and $TiCl_4$, 1:1. The amount of catalyst is 3 grams per hundred of isoprene. The polymerization is complete in 4 hours at 18°C. and is stopped with methanol, after which the *cis*-1,4-polyisoprene is recovered from the mixture (*2*).

Finely divided Li metal was also a catalyst at 30°–40°C although there was an induction period of 1–12 hours, which presumably was the time required to form the true catalyst (*64*). With *n*-butyllithium in benzene or heptane, a precipitate formed and *cis*-1,4-polyisoprene was the predominent product. However there was no induction period (*37, 47*). With *n*-butyllithium in ethyl ether or tetrahydrofuran, no precipitate formed, i.e., the system was homogeneous but the product was mainly 3,4-polyisoprene instead of *cis*-1,4-polyisoprene.

Interestingly, the AlR_3 + TiI_4 catalyst which gave *cis*-1,4-polybutadiene is apparently not active for polymerizing isoprene (*59*).

E. Trans-1,4-polyisoprene

The catalyst is an AlR_3–VCl_3–$Ti(OR)_4$ combination. Prepare an inert support, e.g., kaolin, with about twenty-five layers of VCl_3 on it by suspending the support in dry benzene and adding the required amount of VCl_4.

Warm to decompose the VCl_4 into $VCl_3 + Cl_2$ and when the decomposition is complete, filter (dry box), wash with dry benzene, and then evaporate the residual benzene. Store in an inert atmosphere.

The polymerization system is as follows: To VCl_3 + support, (70 mg of VCl_3,) add $Al(iBu)_3$ (Al:V 20:1); add tetra-2-ethylbutyl titanate (V:Ti 2:1); add benzene 120 gm; and isoprene 100 gm. Polymerize 1.5 hours at 50°C, then kill with methanol and isolate the product (*40*).

Phenylsodium in tetrahydrofuran (a homogeneous catalyst) also gives *trans*-1,4-polyisoprene (*47*) at 40°C, 1 mole % catalyst based on isoprene, and two and a half times as much solvent as monomer.

F. 3,4-Polyisoprene

Using 1 mole % *n*-butyllithium in ethyl ether or tetrahydrofuran (a homogeneous catalyst) at 30°C gave 19% 3,4-polyisoprene. The ratio of solvent to isoprene was 2.5:1. Phenyl- or benzylsodium in *n*-heptane (a heterogeneous catalyst) gave 70–85% 3,4-polyisoprene at 40°C (*29*, *47*).

Li metal in tetrahydrofuran gave 3,4- along with some 1,2-polyisoprene (*34*).

G. Catalyst Suppliers

Climax Molybdenum Co.
Molybdenum pentachloride.

E. I. du Pont de Nemours and Co., Inc.
Alkyl titanates.

Ethyl Corporation
Aluminum alkyls.
alkylaluminum chlorides.

Foote Mineral Co.
Lithium metal.
n-Butyllithium in solution.

Lithium Corporation of America, Inc.
Lithium metal.
n-Butyllithium in solution.

Mackenzie Chemical Works Inc.
Titanium acetylacetonates.

Stauffer Chemical Company
Aluminum alkyls.
Alkylaluminum chlorides.
Titanium tetrachloride.
Titanium trichlorides.
Vanadium trichloride.
Vanadium tetrachloride.
Vanadyl chloride.

All of these materials present certain hazards; the user should know about them and take adequate precautions.

XI. Acetylene to Vinylacetylene

Acetylene is dimerized to vinylacetylene as an intermediate for Neoprene rubber. The author is not aware that the conditions for the commercial process have been published. It is based on Father Nieuwland's (*53*) research on polymerizing C_2H_2 by cuprous chloride solutions. Apparently, the pressure is a little above atmospheric and the temperature around 50°–60°C. Divinylacetylene (a trimer), tetramers, and higher polymers are formed at the same time. These can absorb oxygen from the air readily to form potentially explosive products.

A. Poisons

CO, O_2, H_2S, and other substances that combine with cuprous solutions interfere with the action of the catalyst.

B. Catalyst Supplier

The catalyst is made by the user. One composition (*16*) is: 30–60 gm of Cu_2Cl_2 + 15–35 gm of KCl + 15–45 gm of H_2O made up to a total of 100 gm.

REFERENCES

1. Acciarri, J. A., Carter, W. B., and Flynt, K., *Chem. Eng. Progr.* **58,** No. 6, 85 (1962)
2. Adams, H. E., Stearns, R. S., Smith, W. A., and Binder, J. L., *Ind. Eng. Chem.* **50,** 1507 (1958).
3. Anhorn, V. J., *Proc. 7th World Petrol. Congr.* (1967) Mexico City, Elsevier Pub. Co. Ltd., Barking, Essex, England (1967) **5,** 325 (1967).
4. Anonymous, *Hydrocarbon Process. Petrol. Refiner* **43,** No. 9, 183–184 (1964).
5. Anonymous, *Hydrocarbon Process. Petrol. Refiner* **44,** No. 11, 261 (1965); *Hydrocarbon Process.* **46,** No. 11, 219 (1967).

6. Anonymous, *Hydrocarbon Process.* **46,** No. 11, 216 (1967).
7. Anonymous, *Hydrocarbon Process.* **46,** No. 11, 224 (1967).
8. Anonymous, *Hydrocarbon Process.* **46,** No. 11, 222 (1967).
9. Anonymous, *Chem. Eng. News* **46,** No. 24, 76 (1968).
10. Anonymous, *Chem. Eng. News* **46,** No. 24, 37 (1968).
11. Banks, R. L., and Bailey, G. C., *Ind. Eng. Chem. Prod. Res. Develop.* **3,** 170 (1964).
12. Banks, R. L., Bailey, G. C., Heckelsberg, L. F., and Pennella, F., "Olefin Disproportionation and Catalysts." Meeting of the Catalysis Society, Atlantic City, N.J., Feb. 20–22, 1969.
13. Bogdanovic, B., and Wilke, G., *Proc. 7th World Petrol. Congr.* (1967) Mexico City, Elsevier Pub. Co. Ltd., Barking, Essex, England (1967) **5,** 351 (1967).
14. Bosmajian, G., Burks, R. E., Jr., Feazel, C. E., and Newcombe, J., *Ind. Eng. Chem. Prod. Res. Develop.* **3,** 117 (1964).
15. Bradshaw, C. P. C., Howman, E. J., and Turner, L., *J. Catalysis* **7,** 269 (1967).
16. Carter, A. S., and Johnson, F. W., U.S. Patent 2,200,057 (May 7, 1940).
17. Clark, A., and Bailey, G. C., *J. Catalysis* **2,** 241 (1963).
18. Clark, A., Hogan, J. P., Banks, R. L., and Lanning, W. C., *Ind. Eng. Chem.* **48,** 1152 (1956).
19. Clark, A., Finch, J. N., and Ashe, B. H., *Proc. 3rd Congr. Catalysis, Amsterdam* **2,** 1010 (1964).
20. Crain, D. L., *J. Catalysis* **13,** 110 (1969).
21. Engel, E. F., *Rubber Plastics Age* **45,** 1499 (1964).
22. Feldman, J., Frampton, O., Saffer, B., and Thomas, M., *Am. Chem. Soc. Div. Petrol. Chem. Preprints* **9,** No. 4, A-55 (1964).
23. Field, E., and Feller, M., *Ind. Eng. Chem.* **49,** 1883 (1957).
24. Friedlander, H. N., and Pramuck, E. M. B., U.S. Patent 3,173,904 (March 16, 1965).
25. Friedlander, H. N., and Resnick, W., *Advan. Petrol. Chem. Refining* **1,** 561 (1958).
26. Gadding, E. K., Fisher, B. S., and Collette, J. W., *Am. Chem. Soc. Div. Petrol. Chem. Preprints* **6,** No. 4, A-139 (1961).
27. Gaylord, N. G., and Mark, H. F., "Linear and Stereoregular Addition Polymers," p. 275. Wiley (Interscience), New York, 1959.
28. Gaylord, N. G., and Mark, H. F., *Am. Chem. Soc. Div. Petrol. Chem. Preprints* **6,** No. 2, C-49 (1961); *Makromol. Chem.* **44–46,** 448 (1961).
29. Gibbs, C. F., Horne, S. E., Jr., Macey, J. H., and Tucker, H., *Rubber World* **136,** 705 (1957).
30. Gippin, M., *Ind. Eng. Chem. Prod. Res. Develop.* **1,** 32 (1962); **4,** 160 (1965).
31. Hambling, J. K., Dichlian, M. G., and Spencer, J. G., *Am. Chem. Soc. Div. Petrol. Chem. Preprints* **9,** No. 3, 101 (1964).
32. Heckelsberg, L. F., Banks, R. L., and Bailey, G. C., *Ind. Eng. Chem. Prod. Res. Develop.* **7,** 29 (1968).
33. Heckelsberg, L. F., Banks, R. L., and Bailey, G. C., *J. Catalysis* **13,** 99 (1969).
34. Hsieh, H., Kelley, D. J., and Tobolsky, A. V., *J. Polymer Sci.* **26,** 240 (1957).
35. Jones, E. K., *Advan. Catalysis* **8,** 219 (1956).
36. Kambara, S., Yamazaki, N., and Hatano, M., *Am. Chem. Soc. Div. Petrol. Chem. Preprints* **9,** No. 4, A-23 (1964).
37. Kamienski, C. W., *Ind. Eng. Chem.* **57,** No. 1, 38 (1965).
38. Kelly, R. J., Garner, H. K., Haxo, H. E., and Bingham, W. R., *Ind. Eng. Chem. Prod. Res. Develop.* **1,** 210 (1962).
39. Kennel, W. E., Field, E., and Spiller, C. A., Jr., U.S. Patent 2,973,350 (Feb. 28, 1961).

40. Lasky, J. S., Garner, H. K., and Ewart, R. H., *Ind. Eng. Chem. Prod. Res. Develop.* **1,** 82 (1962).
41. Lehr, M. H., *Rubber Chem. Technol.* **36,** 1571 (1963).
42. Logan, R. S., and Banks, R. L., *Oil Gas J.* **66,** No. 21, 131 (1968).
43. Logan, R. S., and Banks, R. L., *Hydrocarbon Process.* **47,** No. 6, 135 (1968).
44. McKetta, J. J., Jr., *Advan. Petrol. Chem. Refining* **5,** 334 (1962).
45. McMahon, J. F., Bednars, C., and Solomon, E., *Advan. Petrol. Chem. Refining* **7,** 284 (1963).
46. Moore, R. N., Franz, R. A., and Spillane, L. J., *Am. Chem. Soc. Div. Petrol. Chem. Preprints* **14,** No. 1, 118 (1969).
47. Morita, H., and Tobolsky, *J. Am. Chem. Soc.* **79,** 5853 (1957).
48. Mueller, H., Wittenberg, D., Seibt, H., and Scharf, E., *Angew. Chem.* **77,** 318 (1965).
49. Natta, G., *J. Polymer Sci.* **34,** 21 (1959).
50. Natta, G., *Actes Congr. Intern. Catalyse, 2ᵉ, Paris, 1960* p. 39 (1961).
51. Natta, G., and Pasquon, I., *Advan. Catalysis* **11,** 1 (1959).
52. Natta, G., Crespi, G., Valvassori, A., and Sartori, G., *Rubber Chem. Technol.* **36,** 1583 (1963).
53. Nieuwland, J. A., Calcott, W. S., Downing, F. B., and Carter, A. S., *J. Am. Chem. Soc.* **53,** 4197 (1931).
54. Olechowski, J. R., McAlister, C. G., and Clark, R. F., *Inorg. Chem.* **4,** 246 (1965).
55. Ono, I., and Kihara, K., *Hydrocarbon Process.* **46,** No. 8, 147 (1967).
56. Otsuka, S., and Kawakami, M., *Am. Chem. Soc. Div. Petrol. Chem. Preprints* **9,** No. 4, A-33 (1964).
57. d'Ouville, E. L., *in* "Polythene" (A. Renfrow and P. Morgan, eds.), 2nd Ed., p. 35. Wiley (Interscience), New York, 1960.
58. Peters, E. F., Zletz, A., and Evering, B. L., *Ind. Eng. Chem.* **49,** 1879 (1957).
59. Phillips Petroleum Co., Belgian Patent 551,851 (October 17, 1956).
60. Pritzker, G. K., *Petroleum Proces.* **1,** 58 (1946).
61. Reed, H. W. B., *J. Chem. Soc.* p. 1931 (1954).
62. Schrauzer, G. N., *J. Am. Chem. Soc.* **81,** 5310 (1959).
63. Schrauzer, G. N., *J. Am. Chem. Soc.* **82,** 1008 (1960).
64. Stavely, F. W., *Ind. Eng. Chem.* **48,** 778 (1956).
65. Thomas, R. M., Lightbown, I. E., Sparks, W. J., Frolich, P. K., and Murphree, E. V., *Ind. Eng. Chem.* **32,** 1282 (1940).
66. Ueda, K., Oonishi, A., Yoshimoto, T., Hosono, J., Maeda, K., and Matsumoto, T., *Am. Chem. Soc. Div. Petrol. Chem. Preprints* **9,** No. 4, A-47 (1964).
67. Wilke, G., *Angew. Chem.* **69,** 387 (1957).
68. Wilkes, J. B., *Proc. 7th World Petrol. Congr.* (1967) Mexico City, Elsevier Pub. Co. Ltd., Barking, Essex, England (1967). **5,** 299 (1967).
69. Ziegler, K., *Angew. Chem.* **64,** 323 (1952).
70. Ziegler, K., *Angew. Chem.* **72,** 829 (1960).
71. Ziegler, K., Holzkamp, E., Breil, H., and Martin, H., *Angew. Chem.* **67,** 541 (1955).

CHAPTER 9

ALKYLATION

This chapter will deal with adding an olefin hydrocarbon to another molecule so that the net effect is the addition of an alkyl group to the molecule. The alkylation of isobutane with butylenes using concentrated sulfuric acid or liquid anhydrous HF as the catalyst is practiced on a huge scale by the petroleum industry to make highly branched paraffin hydrocarbons which are excellent antiknock components of motor fuel. The techniques devised to control this process illustrate many of the principles of catalysis so that more than the usual amount of detail will be given.

I. Alkylation of Isobutane with Butylenes

In establishing the chemistry of this alkylation, Schmerling (*26–28*) showed that it could be regarded as olefin polymerization by an acidic catalyst with a very fast hydrogen transfer. The acid catalyzed polymerization, e.g., of isobutylene by a carbonium ion mechanism, can be looked on as follows:

$$i\text{-}C_4H_8 + H^+ \rightarrow \begin{array}{ccc} & & C \\ & & | \\ C & - & C^+ \\ & & | \\ & & C \end{array} \quad \text{(H atoms omitted)}$$

This adds to a second olefin:

$$\begin{array}{ccc} & & C \\ & & | \\ C & - & C^+ \\ & & | \\ & & C \end{array} + H_2C{=}C(CH_3)_2 \rightarrow \begin{array}{ccc} & & C \\ & & | \\ C & - & C{-}CH_2\overset{}{C}(CH_3)_2^{+} \\ & & | \quad\quad (A) \\ & & C \end{array} \qquad (4)$$

and this loses H^+ to give the olefin polymer. Writing intermediate (A) as R^+, the hydrogen transfer with isobutane can be written:

$$R^+ + (CH_3)_3CH \rightarrow (CH_3)_3C^+ + \underset{(B)}{RH} \tag{5}$$

This is the hydrogen transfer that characterizes this reaction and this is the reaction all operating conditions are designed to favor. $(CH_3)_3C^+$ is combined with the acidic catalyst and is the compound with the catalyst discussed in Chapter 1. Intermediate (A) is the compound of the product with the catalyst. (B) is the alkylate.

There is also a hydrogen transfer reaction between olefins competing with Eq. (5) which can be represented:

$$i\text{-}C_4H_8 + (CH_3)_3C^+ \rightarrow (CH_3)_3CH + \frac{1}{x}\underset{(C)}{(C_4H_7)_x^+} \tag{6}$$

This reaction is to be minimized because it unnecessarily consumes olefins and because $(C_4H_7)_x^+$ is basic compared to the acid, is soluble in the acid catalyst, and causes acid consumption or loss.

These two reactions and the following facts define the problem.

(1) The catalysts are liquids.

(2) One of the reactants, isobutane, is not very soluble in the catalysts (ca. 0.1% in H_2SO_4 and ca. 3% in HF).

(3) The olefins are soluble in the catalysts and tend to undergo Reaction (6).

(4) Reaction (5) is catalyzed only by strong acid, i.e., 89–98% H_2SO_4 or HF. Weaker acids can cause Reaction (4) but not (5). Thus, if the H_2SO_4 or HF gets too dilute, "run away" polymerization takes place instead of alkylation.

(5) Reaction (4) is reversible; low temperatures favor it going to the right as written. Higher temperatures favor cracking or depolymerization.

In the laboratory or in the plant, continuously stirred reactors (Chapter 1) are used. Some of the techniques used to provide the intense mixing desired to keep an adequate supply of isobutane in the acid as well as to keep the olefin concentration low and so favor Reaction (4) are as follows.

(1) Mixing intensive enough to form an emulsion. An emulsion of hydrocarbon in acid gives better results than the reverse. Acid : hydrocarbon ratios of 1 : 1–3 : 1 or higher favor this.

(2) Injecting the acid and hydrocarbons into a centrifugal pump to provide mixing.

(3) Injecting the acid and hydrocarbons via a jet and eductor into a narrow tubular reactor under turbulent flow velocities, i.e., Reynolds numbers of 10,000 or higher (*1*).

(4) Specially designed mixers with internal cooling coils (*6*).

(5) Several intensive mixing stages in series in a single vessel with appropriate baffles between the stages (*7*).

(6) The isobutane:olefin ratio in the feed is kept as high as practicable, i.e., at least 5:1. By multipoint injection of the stream containing the olefin, local internal isobutane to olefin ratios may approach 1000.

Even with all these tricks, some of Reaction (6) takes place and the undesired product dissolves in the acid and dilutes it. To a very limited extent this is beneficial in that it increases the solubility of isobutane in the acid phase. As the concentration of (C) increases, the detrimental effect soon overcomes the benefit. Ultimately, the acid must be removed. This is accomplished by removing some of the acid and fortifying the rest with stronger acid. With HF the acid is redistilled and (C) is left as a thick dark brown oil. A similar oil can be obtained by diluting the used sulfuric acid with water. These oils are complex mixtures, containing conjugated cyclopentadienes with side chains (*10*, *15*). The net result of this reaction, along with some water that inevitably gets into the acid, is that sulfuric acid consumption is about 60 gm liter^{-1} of the product (*18*) and must be reworked in a sulfuric acid plant. Fresh acid to replenish withdrawal is usually 99.3% H_2SO_4. At first it might seem that even stronger acid would be worthwhile to fortify the remaining acid, but the free SO_3 in the acid causes undesired side reactions.

Since HF is recovered in the plant by distillation, the consumption is much less than with sulfuric acid, about 1 gm liter^{-1} of alkylate. Most of this is rejected as the azeotrope with water that results from water that has gotten into the plant (*20*).

As one reduces the amount of Reaction (6), the yield of product goes up and the acid consumption goes down, as would be expected. However, the antiknock quality of the product goes up too, so there is a great potential technical importance to reducing Reaction (6) (*24*).

As might be expected, there are some problems in operating with HF. It is remarkable, however, that the amount of corrosion is relatively small except where mixtures of HF and water occur. In general, Monel metal has been found to be excellent in HF service even where aqueous HF is present. It goes without saying that silicious materials must be avoided.

With sulfuric acid, the operating temperature is about 10°C. Much lower temperatures are not practical because the viscosity of the acid goes up rapidly as the temperature goes down and this causes the power requirements for mixing to rise exponentially. At much higher temperatures, the oxidizing ability of sulfuric acid becomes noticable and causes undesired effects.

With HF, the operating temperature is somewhat higher, around 25°–35°C. This is not only because HF does not oxidize like sulfuric acid. In terms of "catalyst activity," it might be said that HF is a slightly less active catalyst and requires a higher temperature to provide comparable rates of reaction. This is not a disadvantage because the refrigeration costs for cooling and removing the 20 kcal mole^{-1} heat of reaction can be less. The chemistry of isobutane alkylation with butylenes is far more complicated than indicated by Reactions (4)–(6) (*26–28*). One would expect all the product to be C_8 paraffins, but this in not the case (*9*):

Paraffin	i-C_5	i-C_6	i-C_7	i-C_8	i C_9
Volume % in product	8.4	5.9	6.5	73.1	6.1

As illustrated in Table III, the separate butylenes give C_8 paraffins that do not differ from each other as much as one might at first expect (*31*). Commercially, mixed butylenes are used.

TABLE III

COMPOSITION OF C_8 ALKYLATE FROM DIFFERENT BUTYLENES

Product	1-C_4H_8 (%)	2-C_4H_8 (%)	i-C_4H_8 (%)	Mixed C_4H_8 (%)
2,2,4-Trimethylpentane	41.8	45.2	49.6	43.3
2,2,3-Trimethylpentane	3.4	3.2	3.4	2.2
2,3,3-Trimethylpentane	22.2	23.6	19.1	22.1
2,3,4-Trimethylpentane	17.3	16.4	12.7	20.3
	84.7	88.4	84.8	87.9
2,3-Dimethylhexane	4.8	3.0	3.2	3.6
2,4-Dimethylhexane	5.0	3.6	5.0	3.8
2,5-Dimethylhexane	5.0	4.4	7.0	4.6
3,4-Dimethylhexane	0.5	0.6	—	0.1
	15.3	11.6	15.2	12.1

It should be noted that n-butane or other n-paraffins are not alkylated by olefins and do not react. In practice, the feed always contains some n-butane and this must be separated by distillation before the isobutane is recycled.

It is also to be noted that weaker acids, e.g., phosphoric acid, although capable of olefin polymerization, do not cause alkylation of paraffins. Stronger acids, e.g., $AlCl_3 + HCl$, are catalysts but more expensive.

II. Isobutane Plus Propylene and Pentenes

Isobutane is alkylated by propylene, pentenes, and even higher olefins with the same catalysts (*22*). The reaction conditions are the same primarily because these olefins are usually mixed with butylenes and processed in an existing plant. Invariably, acid consumption goes up [more Reaction (6)] and the antiknock quality of the product declines (*24*). About 10% of the propylene reacts with isobutane in a series of reactions equivalent to:

$$C_3H_6 + i\text{-}C_4H_{10} \rightarrow C_3H_8 + i\text{-}C_4H_8$$

This increases isobutane consumption and gives more C_8 paraffins and less C_7 paraffins in the product than would be expected.

III. Alkylation with Other Saturated Hydrocarbons Plus Olefins

Just looking at other hydrocarbons containing a tertiary hydrogen, one might expect them to alkylate too. However when one looks at the products, (Table III), all of them are hydrocarbons containing at least one tertiary hydrogen, so there must be some limit. A secondary hydrogen on a carbon atom adjacent to the carbon with the tertiary hydrogen atom also comes into play (note that isobutane has only primary hydrogen atoms on the carbon atoms adjacent to the carbon atom with the tertiary hydrogen and in this respect is almost unique). The carbonium ion from isopentane for example tends to lose a secondary hydrogen and form an olefin hydrocarbon:

$$CH_3CH_2\overset{\overset{CH_3}{|}}{\underset{\underset{CH_3}{|}}{C^+}} \rightarrow CH_3CH{=}C(CH_3)_2 + H^+$$

This olefin can react with isopentane to give a C_{10} paraffin alkylate, irrespective of the original olefin. The same thing can happen with methylcyclopentane. These, as well as the rearranged type of products noted in Table III, complicate the products from alkylating these other saturated hydrocarbons (*17*).

Without trying to specify the reaction products, the relative reactivities of a series of saturated hydrocarbons are methylcyclopentane > 2-methylpentane > isopentane > isobutane > methylcyclohexane (*29*).

Just because the trimethylpentanes are formed in sulfuric acid alkylation does not mean that these hydrocarbons are stable to the further action of concentrated sulfuric acid. With 98.5% H_2SO_4 at 10°C, they disproportionate to larger and smaller paraffin hydrocarbons, i.e., $>C_8$ and $<C_8$, with minor amounts of isomerization (*21*).

IV. Isobutane Plus Ethylene

Ethylene does not alkylate isobutane in the presence of H_2SO_4 or HF to a useful extent. The ethylene tends to form esters instead. If BF_3 is added to the HF, then alkylation takes place to give primarily 2,3-dimethylbutane, a useful high antiknock component of motor fuel. 2,2-Dimethylbutane is the expected product but this isomerizes before being released from the catalyst. Practically, $AlCl_3 + HCl$ is the more useful catalyst (*2*, *8*). Actually, the catalyst is a liquid hydrocarbon–$AlCl_3$–HCl complex, with the HCl being generated by the controlled addition of water to the anhydrous $AlCl_3$ system.

Two stirred reactors are operated in series with separate ethylene injection to each reactor. The reaction temperature is about 50°C with a 2°C temperature rise permitted in each reactor (the reaction is strongly exothermic). Cooled hydrocarbon from the second reactor is recycled to the first reactor for temperature control. Liquid catalyst from the second reactor also is recycled to the first. Fresh $AlCl_3$ is introduced into the first reactor by passing isobutane through a bed of anhydrous $AlCl_3$ to dissolve the needed amount. As with the other olefins, the isobutane:olefin ratio is large: 5:1–15:1 or larger.

There is a thermal (450°–500°C), i.e., noncatalytic, process for the alkylation of isobutane with ethylene to give mainly 2,2-dimethylbutane (*13*). This free radical process does not operate to give useful products with higher olefin hydrocarbons.

V. Catalyst Poisons

In all these strongly acidic catalyst systems, anything that reduces acidity must be regarded as a poison. Water is the most prevalent poison, so catalyst consumption can be reduced by drying the feed and recycle hydrocarbons. Molecular sieves are being used more and more as drying agents.

In butylene alkylation, butadiene in the butylenes acts as a poison for sulfuric acid because it does not produce alkylate. Instead, it dissolves in the acid so that 1 kg of butadiene "consumes" 15–20 kg of acid. In HF alkylation, a similar effect takes place and the butadiene increases the load on the HF redistillation tower.

Ammonia brought in with propylene obviously acts as a poison.

VI. Alkylation of Aromatic Hydrocarbons

There are three major commercial demands for the alkylation of aromatic hydrocarbons:

(1) To produce ethylbenzene as an intermediate for styrene (see Chapter 6).

(2) To make cumene (isopropylbenzene) as an intermediate for producing phenol with acetone as a byproduct.

(3) To produce long chain (C_{12}–C_{16}) alkylated benzenes to be sulfonated to make synthetic detergents.

In making ethylbenzene, anhydrous $AlCl_3$ + HCl is probably the most widely used catalyst. The reaction between benzene and ethylene is exothermic, so cooling is used to limit the temperature to a maximum of 95°C. The benzene:ethylene ratio is always greater than 1, and usually 5 or more, to reduce diethylbenzene formation. Most of the diethylbenzene formed is recycled to the reactor (*12*). The catalyst consumption is about 10 kg ton^{-1} of ethylbenzene.

A solid alumina–boron–fluoride catalyst is also available for fixed bed alkylation of benzene with ethylene (*5*). Universal Oil Products Co. supplies the catalyst: Type A–2.

To produce cumene, a mixture of benzene and a refinery stream containing propylene and propane is passed over solid phosphoric acid catalyst (a specially prepared combination of phosphoric acid and kieselguhr) at a space velocity of about 2.5 $V_l V_c^{-1}$ hr^{-1} at a pressure between 25 and 70 atm and a temperature in the range of 180°–225°C (*4, 16*). During use, the catalyst tends to lose water and become inactive so that small quantities of water or isopropyl alcohol ($<0.1\%$ of feed) are added to the reaction mixture to maintain the optimum degree of hydration. Too much water causes the catalyst to get soft and the pellets to collapse, thus increasing the pressure drop in the bed (*23*).

Similarly, the "solid phosphoric acid" catalysts can be used for alkylating benzene with other olefins, such as butylenes or even cyclohexene. Other aromatics may be used, such as toluene and xylenes. Catalysts other than the "solid phosphoric acid" may be used, e.g., sulfuric acid, anhydrous HF, liquid or anhydrous $AlCl_3$. These may be more convenient in the laboratory but on a large scale the "solid phosphoric acid" catalysts are usually more convenient and more economical.

A. Poisons

Organic nitrogen compounds neutralize the acid and reduce effectiveness. Oxygen as low as 20 ppm causes heavy resinous materials to form on the catalyst. Butadiene in a butylene feed also causes the formation of resinous or tarry materials on the catalyst.

B. Suppliers of Proven Catalysts

Both Catalysts and Chemicals, Inc. and Universal Oil Products Co. supply solid phosphoric acid catalysts, and both are phosphoric acids combined with kieselguhr so that the product is solid, both at room and operating temperatures.

C. Detergent Alkylate

Detergent alkylate is made by alkylating benzene with C_{12}–C_{16} olefins using $AlCl_3$, H_2SO_4, or liquid HF as catalysts (*3, 14, 30*). Suitable, but not necessarily optimum, conditions are as follows.

$AlCl_3$: 55°–60°C, 3–4 moles of benzene to olefin, 25 moles of olefin per mole of $AlCl_3$. The reaction time is determined as much by heat removal rate as by reaction rate.

H_2SO_4: 30°C, 4–5 moles of benzene to olefin, 2 volumes of olefin per volume of 96–98% H_2SO_4.

HF: 0°C, 4–9 moles of benzene to olefin, 2–3 kilograms of olefin per kilogram of HF.

VII. Transalkylation of Alkylaromatic Hydrocarbons

In this reaction, an alkyl group moves from one aromatic molecule to another. For example, toluene yields benzene and xylene. The reaction is reversible. The author knows that several companies have given serious consideration to the operation of this type of reaction as a commercial process, but he is not aware of one operating except incidental to another process (e.g., recycling diethylbenzene in making ethylbenzene). The reaction is closely related to the alkylation reactions described above as well as the isomerization of alkyl aromatics (see Chapter 3) for the same catalysts are used for all three reactions. HF plus BF_3 and the acidic catalysts used for cracking petroleum (see Chapter 4) also have been studied.

Interestingly, a mixture of ethylbenzene and *m*- or *p*-xylene with HF + BF_3 at 20°–25°C gave benzene and diethylbenzene but no ethylxylenes (*19*). At a higher temperature, 60°C, all xylenes gave 1,3-dimethyl-5-ethylbenzene, i.e., both isomerization and transalkylation occurred.

This example makes it clear that the operating conditions for the reaction depend on the hydrocarbon system. They also depend on the acidity of the catalyst. The more acidic catalysts (HF + BF_3, $AlCl_3$, or $AlBr_3$) function at or near room temperature. HF alone requires a somewhat higher temperature, up towards 100°C, and the cracking catalysts even higher temperatures, 250°C or higher.

VIII. Alkylation of Phenol

Boron trifluoride is used for the commercial alkylation of phenol with propylene trimer or diisobutylene for the production of nonyl- and octylphenol, respectively. About 0.5–2% of BF_3 is added to the phenol, and the olefin is then added at such a rate that the reaction temperature is maintained in the 40°–45°C range for octylphenol, and the 55°–60°C range for nonylphenol. Higher temperatures tend to produce dialkylated phenols and other byproducts. After the reaction is complete, the catalyst is destroyed by washing with water and then sodium carbonate, after which the mixture is redistilled to recover the desired product and separate it from excess phenol.

Various acidic solids can also be used; for example, the acid treated montmorillonites, silica–alumina, or molecular sieve cracking catalysts. These have to be dehydrated at about 500°C before use. They are added directly to the phenol as finely divided solids, and the olefin is added at a suitable temperature and at a rate to maintain that temperature. An olefinic gas such as propylene is merely bubbled through the reaction mixture, whereas higher boiling olefins, such as nonene or dodecene, are added and refluxed. High boiling alcohols, such as cyclohexanol or dodecanol, can also be used and the corresponding olefin generated *in situ*.

A. Suppliers of Proven Catalysts

See Chapter 4 for suppliers.

IX. Alkylation of the Side Chains of Alkylaromatic Hydrocarbons

In marked contrast to nuclear alkylations catalyzed by acidic catalysts, alkali metal catalysts catalyze the alkylation of the α-hydrogen atoms of the side chain. Thus, ethylene adds to toluene to give *n*-propylbenzene and then 3-phenyl-pentane, and finally 3-ethyl-3-phenylpentane. The reaction temperature is 150°–200°C, with pressures from atmospheric to 10 atm. Propylene requires a higher temperature with toluene (225°–250°C) and gives isobutylbenzene, while isobutylene and toluene give neopentylbenzene (*25*).

Butadiene at 50°–60°C gives 1-phenyl-3-pentene with toluene and, with aniline, gives 2-butenylaniline (*25*).

The catalysts are prepared *in situ* and are sodium–organosodium products made by reacting the metal with organic molecules, such as anthracene and *o*-chlorotoluene, or by depositing the Na on alumina (*25*).

Interestingly, *n*-butyllithium, when chelated with ditertiary amines, catalyzes the reaction between benzene and ethylene to give a mixture of monoalkylbenzenes containing only an even number of carbon atoms in the side chain. With toluene, the catalyst gives side chains on the monoalkylbenzenes that contain only an odd number of carbon atoms. Thus, the reaction is both an alkylation and a polymerization; or it can be regarded as the telomerization of ethylene. Conditions are 70°–110°C and 20–35 atm pressure (*11*).

REFERENCES

1. Albright, L. F., *Chem. Eng.* **73,** No. 19, p. 205 (1966).
2. Alden, R. C., Frey, F. E., Hepp, H. T., and MacReynolds, L. A., *Oil Gas J.* **44,** No. 40, 70 (1946).
3. Anonymous, *Hydrocarbon Process. Petrol. Refiner* **42,** No. 11, 142 (1963).
4. Anonymous, *Hydrocarbon Process. Petrol. Refiner* **44,** No. 11, 248 (1965).
5. Anonymous, *Hydrocarbon Process.* **46,** No. 11, 144 (1967).
6. Anonymous, *Hydrocarbon Process.* **47,** No. 9, 166 (1968).
7. Anonymous, *Hydrocarbon Process.* **47,** No. 9, 164 (1968).
8. Anonymous, *Hydrocarbon Process.* **47,** No. 9, 165 (1968).
9. Cerato, C. C., Sun Oil Company, personal communication, 1967.
10. Deno, N. C., Boyd, D. B., Hodge, J. D., Pittman, C. U., Jr., and Turner, J. O., *J. Am. Chem. Soc.* **86,** 1745 (1964).
11. Eberhardt, G. G., and Butte, W. A., Jr., *J. Org. Chem.* **29,** 2928 (1964).
12. Faith, W. L., Keyes, D. B., and Clark, R. L., "Industrial Chemistry," 2nd Ed., p. 730. Wiley, New York, 1957.
13. Frey, F. E., and Hepp, H. T., *Ind. Eng. Chem.* **28,** 1439 (1936).
14. Holst, E. H., Startz, A. J., Zoller, J. R., and Tessieri, J. E., *Ind. Eng. Chem. Prod. Res. Develop.* **1,** 120 (1962).
15. Hughes, E. C., Stevens, D. G., and Veatch, F., *Ind. Eng. Chem.* **43,** 1447 (1951).
16. Jones, E. K., *Advan. Catalysis* **8,** 233 (1956); **10,** 184 (1958).
17. Kennedy, R. M., *Catalysis* **6,** 32 (1958).
18. Kobe, K. A., and McKetta, J. J., Jr., *Advan. Petrol. Chem. Refining* **2,** 314 (1959).
19. McCaulay, D. A., Hoff, M. C., Stein, N., Couper, A. S., and Lien, A. P., *J. Am. Chem. Soc.* **79,** 5808 (1957).
20. McKetta, J. J., Jr., *Advan. Petrol. Chem. Refining* **3,** 278 (1960).
21. Morton, F., and Richards, A. R., *J. Inst. Petrol.* **34,** 133 (1948).
22. Mrstik, A. V., Smith, K. A., and Pinkerton, R. D., *Advan. Chem. Ser.* **5,** 97 (1951).
23. Nelson, W. L., *Oil Gas J.* **63,** No. 35, 104 (1965).
24. Panel of Experts, *Oil Gas J.* **64,** No. 35, 124 (1966); **66,** No. 9, 96 (1968).
25. Pines, H., and Schaap, L. A., *Advan. Catalysis* **12,** 117 (1960).
26. Schmerling, L., *J. Am. Chem. Soc.* **66,** 1422 (1944); **67,** 1152, 1778 (1945); **68,** 275 (1946).
27. Schmerling, L., *Ind. Eng. Chem.* **40,** 1072 (1948).
28. Schmerling, L., *Chem. Petrol. Hydrocarbons* **3,** 363–407 (1955).
29. Schneider, A., *J. Am. Chem. Soc.* **76,** 4938 (1954).
30. Sharrah, M. L., and Feighner, G. C., *Ind. Eng. Chem.* **46,** 248 (1954).
31. Zimmerman, C. A., Kelly, J. T., and Dean, J. C., *Ind. Eng. Chem. Prod. Res. Develop.* **1,** 124 (1962).

CHAPTER 10

CATALYSTS FOR FREE RADICAL POLYMERIZATIONS

Organic compounds that decompose at relatively low temperatures, yielding free radicals, are used to initiate radical chain-reaction polymerizations. These polymerizations may be homogeneous, to form plastics, or heterogeneous emulsions, e.g., to form polystyrene or polyvinyl chloride. These decomposable initiators are often miscalled catalysts because they have some of the superficial characteristics of true catalysts. They are treated here because catalysts are used with them and because it is advisable to understand their action to use the catalysts with them.

I. Free Radical Polymerizations

Free radical polymerizations are characterized by an induction period during which very little, apparently, happens. After the induction period, there is an exothermic reaction during which polymerization occurs and polymer forms. Finally, as the monomer is exhausted, the exothermic reaction stops and polymerization is complete. These reactions are well-known free radical chain reactions, and the induction period is really the period during which the chain reaction is initiated. During the exothermic period, there is chain branching and, finally, chain termination.

Peroxides and organic peroxygen compounds, as well as certain aliphatic azo compounds, are well known as initiators for these polymerizations. They decompose, liberating free radicals which initiate the polymerization. In other words, they shorten the induction period. The degree of shortening of the induction period is related to the concentration of free radicals which, in turn, is related to the concentration of the initiator and the rate at which it decomposes to give free radicals.

Not all initiators decompose at the same temperature, so there is a range of temperatures at which they initiate polymerization. Also, not all monomers are equally reactive, so there is a problem of matching the initiator to the monomer activity.

It is the purpose of this section to help the reader select an initiator when he needs one in his work. Also, it will give him some measure of the temperature at which various initiators become effective and useful. The temperature at which one-half of the initiator decomposes in 10 hours is available for many of these compounds and is one such measure. Table IV lists some

TABLE IV

TEMPERATURE AT WHICH HALF LIFE IS 10 HOURS FOR VARIOUS INITIATORS

Initiator	$t_{1/2}$ = 10 hrs (°C)
Isopropyl percarbonate	35
2,4-Dichlorobenzoyl peroxide	54
Lauroyl peroxide	62
α,α'-Azobisisobutyronitrile	63
Caprylyl peroxide	63
Acetyl peroxide	69
Benzoyl peroxide	72
p-Chlorobenzoyl peroxide	75
t-Butyl peroxyisobutyrate	79
Hydroxyheptyl peroxide	85
Cyclohexanone peroxide	91
t-Butyl peracetate	102
Di-*t*-butyl diperphthalate	105
t-Butyl perbenzoate	105
Methyl ethyl ketone peroxide	105
Dicumyl peroxide	117
2,5-Dimethylhexane-2,5-dihydroperoxide	154
Cumene hydroperoxide	158
t-Butyl hydroperoxide	172

of the common commercial initiators and their thermal stabilities (the temperature required to decompose one-half of it in 10 hours under selected standard conditions). With this as background, the user can select a catalyst to aid the initiator, if one is needed.

II. Catalysts or Promotors

Some of the initiators can be made more active by catalysts or promotors. These catalyze the decomposition of the peroxy compound so that it occurs at a much lower temperature than it would otherwise. Because of this, *never* mix promoter and peroxy compound, for a violent and possibly explosive reaction can occur. Instead, add the promoter to the monomer and mix thoroughly before adding the peroxy compound. Many promoters have been tried and used. The principal ones are:

(1) Tertiary amines for the diacyl peroxides. Dimethylaniline is good and in some cases dimethyl-*p*-toluidine is enough better to justify the added cost.

(2) Monomer soluble cobalt compounds (usually cobalt naphthenate) are promoters for ketone and aldehyde peroxides.

(3) High molecular weight mercaptans (e.g., lauryl or *t*-dodecyl mercaptan) are the alkyl perester promoters.

These promoters or catalysts can dramatically lower the temperature at which chain branching and rapid polymerization begins. In fact, they are the backbone of the room-temperature-curing polyesters that make plastic boats, automobile bodies, etc., possible.

Many of the initiators are solids that may be relatively hard to dissolve or disperse in the monomer. As a result, the manufacturer usually supplies several grades of the initiator in the form of a paste so that the solid is easily dispersible or dissolved. Also, the manufacturer may supply the solid compound as a concentrated solution in a solvent that does not interfere with the polymerization.

The amount of initiator used is 0.1–1%, based on the monomer. The amount of catalyst or promoter is usually less than the initiator and about 0.05–0.1%. It must be emphasized that these vary widely with the monomer, the initiator, and the catalyst.

III. Safety

These compounds are useful because they are thermally unstable. Some must be shipped and stored below room temperature. Many are powerful oxidizing agents and must be stored away from readily reducible substances. Eyes and skin must be protected. The supplier has detailed directions for the safe handling of his products and is glad to furnish them.

IV. Suppliers of Proven Products

Since many of the initiators are sold as solutions or pastes of varying concentration, and sometimes under trade names, suppliers only are listed below rather than individual products. Moreover, the list in Table IV does not include all commercially available initiators.

Like the catalyst suppliers, the suppliers of initiators often have useful literature on their products which they gladly furnish.

Chemetron-Noury Corp.
E.I. Du Pont De Nemours & Co.
Hercules, Inc.
Lucidol Division, Wallace & Tiernan, Inc.
PPG Industries, Chemical Division
Thompson Chemical Co.
U.S. Peroxygen Corp.

The following supply cobalt catalysts soluble in organic media.

Carlisle Chemical Works, Inc.
Harshaw Chemical Company
MacKenzie Chemical Works, Inc.
Nuodex Division of Tenneco Chemicals, Inc.
Witco Chemical Corporation, Organics Division

CHAPTER 11

SYNTHESIS GAS AND HYDROGEN PRODUCTION

Synthesis gas is a mixture of hydrogen and carbon monoxide used for the Oxo process or making methanol. It is also an intermediate stage in the production of hydrogen. It can be produced by noncatalytic processes, e.g., oxygen and steam + coal, coke, or liquid hydrocarbons. Only the catalytic processes will be treated here. Producing ammonia synthesis gas or reducing gases for metallurgical operations is a closely related process.

I. Synthesis Gas from Gaseous Hydrocarbons

In its simplest form this process can be represented:

$$CH_4 + H_2O \overset{Ni}{\rightleftharpoons} CO + 3\,H_2$$

As used, the process is considerably more complicated. The reaction is reversible, with higher temperatures favoring the $CO + H_2$. Increasing the partial pressure of H_2O (steam) also tends to drive the reaction to the right, so steam:carbon ratios of 2:1–6:1 or higher are used (the steam:carbon ratio is the ratio of the number of moles of steam to the number of carbon atoms in the hydrocarbon). The excess steam has an added benefit in that it reduces the carbon formation on the catalyst. Carbon formation tends to disintegrate some catalysts mechanically.

Excess steam is expensive in that it must be heated to reaction temperature, so the final steam:carbon ratio is a compromise. Special catalysts have been developed that permit low steam:carbon ratios without carbon formation.

Operating pressure is also a compromise. Low pressure favors $CO + H_2$ but most $CO + H_2$ (or H_2) is used at elevated pressures and compressing the gases is costly. The trend has been from 2–5 atm in early plants to 25–40 atm in more recent plants. The ability to operate the process at higher pressures has permitted great reductions in the cost of producing ammonia or methanol (*2*). The details are given in Chapter 14.

The operating temperature is limited primarily by the available metallurgy technology. As of this writing, 1000°C at 30 atm, or possibly higher, is practical. The reaction is strongly endothermic, so the process is conducted in externally heated tubes. It is desirable to supply this heat at as low a temperature as practical, e.g., around 650°C, so highly active catalysts have been developed for this use, but several percent CH_4 remains in the gas. Temperatures to 1000°C or higher are needed to reduce the CH_4 to 0.1–2.0% in the product gas; more rugged, less active catalysts are used at these high temperatures.

The interrelation of the several process variables on the steam–methane equilibrium has been discussed by Wellman and Katell (*7*).

Here it should be pointed out that the steam reforming of methane produces more H_2 in the product gas than needed in the methanol synthesis. The carbon monoxide content can be increased in several ways (*5*), e.g., the steam may be partly replaced by CO_2:

$$CH_4 + CO_2 \overset{Ni}{\rightleftharpoons} 2\,CO + 2\,H_2$$

The same Ni catalysts are effective.

Summarizing, the steam–natural gas reforming process uses nickel on refractory supports as catalysts at temperatures in the range of 650°–1000°C, at pressures of 3–40 atm, with steam (or CO_2):carbon ratios of 2–6 or higher, at a GHSV of 5000–8000 $V_g V_c^{-1}$ hr^{-1}, based on $H_2 + CO$ at room temperature and pressure.

II. Synthesis Gas from Naphthas

Synthesis gas can be produced from steam and naphthas and, since the naphthas are more reactive than natural gas, lower temperatures are used, e.g., 700°–825°C (*1*), although some catalysts operate as low as 550°–650°C. The tendency to form carbon during the reaction is greater with naphthas than with natural gas, so the steam:carbon ratio must be kept above 2:1. Special dense and rugged catalysts have been developed that are resistant to the disruptive effects of coking, and these are used in some reactors in the forepart of the reactor tube. Here the chances for coking are greatest.

A less dense, more reactive catalyst is placed towards the exit part of the reactor tube (*4*).

In summary, operating conditions are: temperature 550°–825°C, pressure 3–40 atm, with steam:carbon ratios of 2:1–6:1, and an LHSV of 1–4 $V_l V_c^{-1}$ hr^{-1} based on the naphtha. The product gas varies with the composition of the naphtha and with the operating conditions. The following gas was produced at 750°C, 8 atm and H_2O:C of 3.3:1: CH_4 4%; H_2 69%; CO 14%; CO_2 12% (*3*).

A. Poisons

Sulfur is a temporary poison, so both natural gas and naphtha reformers are preceded by desulfurizers. As has been mentioned, carbon or coke may physically damage the catalyst. Coke that has not damaged the catalyst can be removed by steam, i.e., by lowering or cutting off the hydrocarbon feed. Heavier coke is burned off with steam + air. Olefins in the feed can cause coke deposits, especially in a natural gas reformer. Arsenic is a permanent poison.

B. Suppliers of Proven Catalysts

Catalysts and Chemicals Inc.

- C11–2 and C11–5: high nickel oxide on calcium aluminate refractory supports as Rashig Rings for reforming light hydrocarbons and hydrogen production.
- C11–2S: contains less nickel but is also on a calcium aluminate support as Rashig Rings. This is for severe service, i.e., high temperature and high pressure.
- C11–7: nickel oxide on a refractory support. For the entrance part of the catalyst bed in naphtha reforming.
- C11–8: also nickel oxide on a refractory support and for the exit part of the catalyst bed in naphtha reforming.

Catalysts Division, Chemetron Corporation

- Girdler G–56 and G–56B: nickel based catalysts that are resistant to physical degradation by carbon formation.

Katalco Corporation

- 22–6(35): nickel oxide–alumina for primary reforming. The ABD is 1.1, the surface area 50–60 m^{2}gm^{-1}, and the pore volume 0.2–0.3 cm^{3}gm^{-1}.
- 54–1: nickel oxide–alumina for secondary reforming. The ABD is 1.0, the surface area 50–60 m^{2}gm^{-1}, and the pore volume 0.2–0.3 cm^{3}gm^{-1}.
- 46–1: nickel oxide–alumina–silica–promoter for naphtha reforming. The

ABD is 1.15, the surface area 50–60 m^2gm^{-1}, and the pore volume 0.2–0.3 cm^3gm^{-1}. Available only to licensees of the process.

46–2: nickel oxide–alumina–silica–promoter for naphtha reforming. The ABD is 1.15, the surface area 50–60 m^2gm^{-1}, and the pore volume 0.2–0.3 cm^3gm^{-1}. Available only to licensees of the process.

In general, the life of steam reforming catalysts is 5 years or longer.

III. Hydrogen Production and the Shift Process

To make hydrogen from synthesis gas, the CO is reacted with steam to make additional hydrogen (*6*):

$$CO + H_2O = H_2 + CO_2$$

The CO_2 is then removed by special solvents. Since hydrocarbons are not converted in this process, the steam reforming process is operated to convert as much of the hydrocarbons as practical, i.e., the CH_4 content should be well under 1%.

The process is operated in adiabatic reaction chambers, and since the process is exothermic, there is a temperature rise in the reactor. However, the reaction is reversible, and higher temperatures favor CO instead of H_2. As a result, the process is conducted in two or even three stages, with intermediate cooling. For the first stage, the gases from the steam reformer are cooled to the operating temperature of the shift reactor, and more steam may be added. In the first stage, the catalysts are iron oxide–chromium oxide compositions; they operate at 315°–485°C, with space velocities (dry inlet gas basis) of 300–4000 $V_gV_c^{-1}$ hr^{-1}, and H_2O:CO ratios of 2–4. Pressure can be as high or low as the rest of the process requires. In general, at a given space velocity with inlet gas measured at atmospheric pressure, the apparent activity increases with pressure due to the longer contact time. 90–95% CO conversion is usual, but this is obviously related to operating temperature. The CO in the product gas is about 1–5%.

If insufficient steam is present, the iron oxide may be reduced to metallic iron, a catalyst for the reaction:

$$2\,CO \rightarrow CO_2 + C$$

This is to be avoided, as the C damages the catalyst. For the same reason, both steam and gas should be free of oil (e.g., compressor or pump oil), for this may carbonize on the catalyst.

After the first stage, the gases are cooled to 175°–350°C, so the lower temperature will favor the conversion of even more of the CO. The more active zinc oxide–copper oxide catalysts with Zn:Cu ratios usually in the range of 0.5 : 1–3 : 1 are used. Again, the pressure is that required by the rest of the operation, and may be 3–40 atm, and the GHSV is 300–4000 $V_g V_c^{-1}$ hr^{-1}, to give a conversion efficiency of 95–99%. Like the first stage, the reaction is conducted in adiabatic catalyst cases. The CO concentration in the product is 0.05–0.5%. After the low temperature shift, the CO_2 is extracted. If the residual CO is a poison for the hydrogenation operation, it is reduced to methane (see Chapter 12) or oxidized to CO_2.

A. Poisons

The iron based catalysts for the first stage are not as susceptible to sulfur poisoning as the nickel catalysts used in the steam reforming step. Thus, synthesis gas made from coke that contains sulfur may contain H_2S and possibly organic sulfur compounds. 500–1000 ppm of H_2S and 20 ppm of organic sulfur may be tolerated by some catalysts. In fact, if COS is in the gas, it is hydrolyzed to CO_2 and H_2S during the "shift" (*8*).

On the other hand, the low temperature ZnO–CuO catalysts are permanently poisoned by sulfur and chlorides. Water used in the reaction should be watched for chloride content.

B. Suppliers of Proven Catalysts

Catalysts and Chemicals Inc.

C12–1: a chromium promoted iron oxide for the high temperature (first stage) shift for use with gases containing less than 20 ppm of sulfur compounds.

C12–3: also a chromium promoted iron oxide, for gases containing more than 20 ppm of sulfur.

Both catalysts are supplied as $\frac{3}{8} \times \frac{1}{4}$ and $\frac{3}{8} \times \frac{3}{16}$ inch tablets.

C18–1: copper oxide–zinc oxide–alumina ($\frac{1}{4} \times \frac{1}{8}$ tablets and $\frac{1}{8}$–$\frac{1}{4}$ inch spheres) for the low temperature (second stage) shift, as well as the improved version, C18–3–02.

Catalysts Division, Chemetron Corporation:

Girdler G–3A: a chromium oxide promoted iron oxide for the high temperature shift.

Girdler G–66: a Cu–Zn–alumina catalyst.

Girdler G–66B: a Cu–Zn catalyst.

The latter two are supplied for the low temperature shift.

Katalco Corporation

15–4: an iron oxide–chromia catalyst for high temperature shift. The ABD is 1.0, the surface area 60–80 m^2gm^{-1}, and the pore volume 0.25–0.27 cm^3gm^{-1}.

52–1: a copper oxide–zinc oxide–alumina catalyst for low temperature shift. The ABD is 1.0, the surface area 60–80 m^2gm^{-1}, and the pore volume 0.2–0.3 cm^3gm^{-1}.

IV. Ammonia Synthesis Gas

Any gaseous hydrocarbons, e.g., methane, left in the hydrogen for ammonia synthesis pass the ammonia catalyst unchanged, and since unconverted gas is recycled, the hydrocarbons build up and reduce the hydrogen partial pressure. In making ammonia synthesis gas, the hydrocarbons are reduced to 0.2–0.5%. In the steam–natural gas reforming step, the primary reformer may be operated so that 5–10% methane is left in the product gas. This is then mixed with controlled quantities of air (nitrogen is needed in the synthesis gas) and the mixture passed over "high temperature" secondary reforming catalysts contained in refractory lined adiabatic catalyst cases. The temperature rises to about 1200°C as a result of the exothermic reaction. Pressure is substantially the same as in the primary reformer. The catalysts are usually NiO on refractory supports, although chromium oxide on a refractory support is used at temperatures in the range of 1200°–1800°C, expecially at the inlet part of the catalyst case, where upsets might result in excessive temperatures that would otherwise damage the catalyst. The high temperatures carries the steam reforming reaction to near completion. The product gas is then cooled and fed to the shift converter after which the CO_2 is removed. Any residual CO is hydrogenated to methane or selectively oxidized to CO_2, which is extracted.

A. Poisons

Sulfur is a temporary poison. Its inhibiting action ceases shortly after the source of sulfur is removed.

B. Suppliers of Proven Catalysts

Catalysts and Chemicals Inc.
C14: a nickel oxide on refractory support.
C15: a chromium oxide base catalyst.

Catalysts Division of Chemetron Corp.
G–31 A: 4.5% Ni on ⅝–1½ inch α-alumina lumps. ABD is 1.6. For secondary reforming.
G–31 B: 7% Ni on ⅝–1½ inch α-alumina lumps. ABD is 1.6. For autothermic reforming.

V. Metallurgical Gases

It should be obvious from the technology summarized above that these methods can be used in various forms for producing a variety of atmospheres used in metallurgical operations. In addition, it is possible to omit the steam and to just use hydrocarbons and air to crack the hydrocarbons and make gases containing N_2, H_2, CO, CO_2, H_2O, and CH_4, with the composition depending on (*a*) the hydrocarbon fed, (*b*) the air:hydrocarbon ratio, and (*c*) the operating temperature. The operation can be conducted so that supplemental (external) heat is required, e.g., at about 1000°C, or in a self-sustaining operation at about 1000°–1200°C. In both cases, catalysts with a low nickel content are used.

A. Supplier of Proven Catalysts

Harshaw Chemical Co.

Ni–0901 S: a 5% Ni (as NiO) on an inert silica–alumina support for use at about 1000°C. The ABD is 1.2 and the surface area 1 m^2gm^{-1}.

Ni–2002 C Spec 103: 5% Ni (as NiO) on one inch inert cubes. This is for use in either type of operation.

Ni–1901 G 2–1 : 3% Ni (as NiO) on an activated alumina for use at 1000°–1200°C. The ABD is 0.85.

REFERENCES

1. Anonymous, *Hydrocarbon Process. Petrol. Refiner* **42,** No. 11, 223 (1963).
2. Anonymous, *Oil Gas J.* **63,** No. 30, 50 (1965).
3. Fox, J. M., and Yarze, J. C., *Ind. Chemist* **40,** No. 2, 65 (1964).
4. Gignier, J. P., and Quibel, J. H., *Hydrocarbon Process. Petrol. Refiner* **44,** No. 3, 153 (1965).
5. Mehta, D. D., *Hydrocarbon Process. Petrol. Refiner* **47,** No. 12, 127 (1968).
6. Moe, J. M., *Chem. Eng. Progr.* **58,** No. 3, 33 (1962).
7. Wellman, P., and Katell, S., *Hydrocarbon Process. Petrol. Refiner* **42,** No. 6, 135 (1963).
8. Wilson, M. W., and Plants, K. D., *Ind. Eng. Chem. Process Design Develop.* **7,** 526 (1968).

CHAPTER 12

GAS PURIFICATION

Industry encounters many gas streams containing small concentrations of undesirable impurities. For pollution control, and for other purposes, there is a great demand for highly selective catalysts that will reach in and convert these impurities into substances that are innocuous or substances that are readily removed. The catalytic art has provided a number of useful solutions to these problems. There is an entire book (7) that is useful in this area, so this chapter provides only a summary along with specific catalysts.

Broadly, most of these processes are required to treat large volumes of gas at or near atmospheric pressure with minimum pressure drop. As a result many of the catalysts are deposited on "open" supports such as wire screens, honeycomb ceramics, ceramic rods, and similar shapes.

I. Removing Traces of Combustible Material from Air

In many industrial operations, small amounts of objectionable odors and combustible materials ("fumes") are left in air and these must be removed lest the plant becomes a public nuisance. Some of the examples are: the air blowing or asphalt in which objectionable aldehydes and odors remain in the offgas; core ovens in foundries where gas and oil vapors are released; the manufacture of formaldehyde where small amounts of combustible materials remain in the offgases; the printing and lithography operations where small amounts of solvents and resins may be in the offgas; the manufacture of phthalic anhydride where small amounts of materials remain in the offgas; from varnish cooking where distinct odors of hot linseed oil and the like remain; from wire coating and enameling ovens where solvent and varnish may remain in the gas; from baking and cooking where odors remain in the gas or air and these may be objectionable over long periods of time. Keeping the air clean in an atomic submarine or removing traces of C_2H_2 from the air to a liquid oxygen plant present similar problems.

Each application usually requires its own solution so the catalyst suppliers also provide the equipment to meet the requirements. In general, provision must be made to heat the catalyst to operating temperature (350°–450°C). If there is enough oxidizable material present, the temperature is maintained or even rises. Gas space velocities in the range of 30,000–150,000 $V_g V_c^{-1}$ hr^{-1} are common.

In the atomic submarine, traces of organic vapors, or gases, or even aerosols of lubricating oil and hydraulic fluids can find their way into the confined atmosphere. To reduce any possible hazard, a continuous stream from the atmosphere is passed over a hopcalite catalyst. Suitable conditions are 350°C, GHSV of 21,000 $V_g V_c^{-1}$ hr^{-1} for air containing 20–125 mg m^{-3} of organic vapors or 15–60 μg liter^{-1} of aerosols (*4, 5*). While this treatment removes most organic vapors, methane is only partially oxidized under these conditions.

Liquid oxygen plants are usually located in industrial or urban areas where the air will contain traces of acetylene. When such air is fed to a liquid air–liquid oxygen plant, the C_2H_2 collects in the liquid oxygen reboiler. Since it is much less volatile than oxygen, it concentrates there and forms a potential hazard. Explosions have wrecked such plants and the C_2H_2–O_2 mixture may have been the cause. It is common practice periodically to withdraw and discard some of the liquid and in this way keep the C_2H_2 concentration down. A turbocompressor for a liquid air plant will discharge the compressed air at about 5–10 atm and 150°C. Under these conditions, hopcalite will oxidize C_2H_2 to CO_2 and water so that the concentration of C_2H_2 is reduced to less than 0.1 ppm (*8*). Adding 1.0% Ag_2O to the 60% MnO_2–40% CuO catalyst lessened its sensitivity to H_2O.

Hopcalite was developed during World War I as a component in gas mask canisters for oxidizing CO to CO_2. It is still used with some gas masks.

A. Suppliers of Proven Catalysts

The Air Correction Division of Universal Oil Products Co.

Pt deposited on high nickel alloy ribbons which are woven into multilayer mats. Packaged units are supplied to meet specific applications.

Catalyst Division of Chemetron Corporation

Girdler G–43: a Pt–Ni on alumina catalyst as ¼ and ⅜ inch tablets. The ABD is 1.

Matthey Bishop, Inc.
Pt deposited on a honeycomb ceramic (Du Pont Torvex) as well as packaged units for specific applications.

Mine Safety Appliance Company
Hopcalite: 60% MnO_2–40% CuO specially prepared for high activity, as well as "catalytic filters" using it.

Oxy-Catalyst, Inc.
Pt on small ceramic rods that are supported at each end by ceramic headers so that a group of them forms a "brick" $3 \times 3\frac{3}{16} \times 5\frac{1}{2}$ inches. The rods are spaced far enough apart so the pressure drop through them is small. The individual bricks are stacked side by side and on top of each other to form the catalyst bed. Oxy-Catalyst also supplies units with Pt on ceramic honeycomb. Packaged equipment using these catalysts is available for specific applications.

II. The Catalytic Muffler (Unleaded Fuels)

The Otto or Diesel cycle internal combustion engines leave small amounts of unburned materials in the exhaust gas as well as some nitrogen oxides. The Otto cycle engine leaves some CO. These pollute the air and in some cases can cause death. Many pieces of equipment, e.g., fork lift trucks, need to be operated in enclosures where these contaminants cannot be tolerated. The regular muffler on such equipment is replaced with a "catalytic muffler" that not only muffles the sound of the engine but also oxidizes the residual combustible matter to tolerable levels. Pt on a refractory (usually ceramic) support is used as the catalyst. Since the Pt is poisoned by lead in fuels containing lead alkyls, unleaded fuels are used. Diesel engines normally operate on unleaded fuels, but the fuel normally supplied for Otto cycle engines (automobile engines) usually contains lead alkyls, so care must be exercised in providing fuels for the engines with catalytic mufflers.

A. Suppliers of Proven Catalysts

Engelhard Industries, Matthey Bishop, Inc., and Oxy-Catalyst, Inc. supply Pt catalysts for this purpose.

III. The Catalytic Muffler (Leaded Fuels)

Most of the common motor fuels contain lead alkyls as antiknock agents. Aliphatic chlorides and bromides (usually ethylene dichloride and ethylene dibromide) are present to help scavenge the lead compounds from the combustion chamber. Organic phosphorous compounds are often added (e.g., 0.1–0.2 stoichiometric theories of P based on the Pb) to retard the fouling of spark plugs by the lead compounds formed on combustion. A few gasolines also contain boron compounds. Finding catalysts for mufflers that maintain activity with such a melange has been a difficult problem that is not completely solved as of this writing. The lead compounds discharged by the engine are solids and the catalyst bed acts a filter and collects much of the solids ultimately to plug the catalyst bed. The lead compounds may or may not act as catalyst poisons, depending on the catalyst. The halides and phosphorous compounds may actually damage the catalyst more than the solid lead compounds.

The catalyst is expected to start functioning at as low a temperature as possible and yet be rugged enough to function and withstand temperatures up to 700°C or higher. To handle around 5 m^3 min^{-1} of gas, with a GHSV around 10,000–30,000 $V_g V_c^{-1}$ hr^{-1}, 5–10 liters of catalyst must be in the catalytic muffler.

The catalyst must be physically strong enough to resist breakage, as a result of many heating and cooling cycles, as well as considerable mechanical vibration. As of this writing it has been agreed that a catalytic muffler that can function for 12,000 miles and still retain a specified activity is acceptable. The catalysts that currently are approaching or meeting this requirement are copper oxide–chromium oxide combinations on activated alumina or copper oxide–vanadium oxide on active alumina. Such catalysts are as of now still under active competitive development, so the manufacturers are understandably reluctant to give details on composition and performance or even to indicate that their catalyst is available commercially. American Cyanamid Company representatives (*9*) kindly furnished the data in Table V

TABLE V

COMPOSITION AND PROPERTIES OF A CATALYST FOR OXIDIZING AUTOMOBILE EXHAUST GASES

V_2O_5 (%)	4–7
CuO (%)	3–7
Pd (%)	0.01–0.015
SiO_2 (%)	5
Al_2O_3	Balance
Surface area (m^2 gm^{-1})	146
Pore volume (cm^3 gm^{-1})	0.72
ABD	0.65
Diameter of particle (in)	0.060
Length of particle (in)	0.11

on one of their Aero-Ban catalysts. This catalyst was used in ten different automobiles, each driven 12,000 miles with six different motor fuels, all except one containing some lead alkyls. Correlations were established from which "averages" could be obtained. The average hydrocarbon emissions increased from 118 ppm at the beginning of the test to 358 ppm at the end of 12,000 miles; CO increased from 0.5 to 1.3%. The greater the lead alkyl content of the motor fuel, the faster the decrease in activity. The surface area of the catalyst also decreased and activity decreased with area but this was not related in a simple way to the lead alkyl content of the motor fuel. Table VI

TABLE VI

CORRELATED RESULTS ON AERO-BAN CATALYST AFTER 12,000 MILES USE

Lead alkyl content of motor fuel (ml gal.$^{-1}$)	Pb on used catalyst (%)	For 100 m^2 gm^{-1} final surface area		For 50 m^2 gm^{-1} final surface area	
		Relative activity[a]	% of fresh activity	Relative activity[a]	% of fresh activity
0	0	461	83	416	75
1	10	435	78	369	66
2	17	411	74	324	58
3	24	391	70	287	52
4	30	371	67	250	45

[a] The relative activity is based on a laboratory test under standard conditions performed on a sample of each catalyst after removing it from the muffler.

summarizes the correlated data. These results are presented not with the idea that they have significance in an absolute sense but rather to indicate the remarkable ability of such catalysts to maintain significant activity under such adverse conditions.

Activated alumina supported copper oxide–chromium oxide gives results essentially comparable with the above results using the vanadium-containing catalyst; however, neither type catalyst is as good as desired for reducing the oxides of nitrogen. So far as the author knows, a practical catalyst for this purpose under these conditions has not been found.

IV. Removing Sulfur Compounds from Producer, Coke Oven, Hydrogen, or Hydrocarbon Gas Streams

There are many cases where removing sulfur compounds from gas streams is necessary. Removal is to be taken literally, i.e., the solid retains the sulfur or sulfur compound. The solid may or may not be regenerable. Both reagents and catalysts for sulfur removal are discussed because sulfur removal is so important for so many catalysts. Catalysts that convert the sulfur from one compound to another but do not remove it are not considered here.

One of the oldest selective catalytic devices for removing trace impurities was applied to coke oven gas where H_2S was to be removed. In one of its versions, this was a simple box containing alkalized bog iron ore and a certain amount of excelsior or wood chips to keep the pressure drop through the bed low. In the presence of oxygen, this purification box catalytically converted hydrogen sulfide to elemental sulfur which remained in the box. Enough oxygen had to be in the gas to effect this conversion. At the same time, a certain amount of sulfur combined with the iron to form noncatalytic iron sulfide (7). After many regenerations enough of the iron ore became converted to inert iron sulfide and sulfate so that the material has to be replaced.

This process works on producer, coke oven, and natural gas. It is being used on natural gas under pressure. Since it can operate at essentially atmospheric temperature, presumably it works on any gas that is not too strongly acidic. If present, the iron oxide removes HCN and NO.

Metallic oxides supported on activated carbon can be used for desulfurizing gases, especially hydrocarbon gases to be used in steam reforming. They are effective for removing alkyl mercaptans and disulfides, thioethers, and thiophenes. They have limited capacity for H_2S and virtually none for COS. They operate at 15°–50°C and atmospheric to 40 atm or higher pressures. Residual total sulfur is 0.1 ppm (by volume) or less. From 50,000 to 200,000 volumes of gas are treated per volume of catalyst before regeneration. The catalysts are used with gases containing up to 50 ppm (by volume) of sulfur.

Zinc oxide, alone or supported, will remove H_2S, alkyl mercaptans, and disulfides, thioethers, and COS from gas streams containing up to 200 mg m^{-3} of sulfur compounds. The operating temperature is 300°–400°C and up to 75 atm pressure or higher. The GHSV is 500–2000 $V_g V_c^{-1}$ hr^{-1} for dry gas or up to 20,000 for wet gas, e.g., as a desulfurizer (guard case) preceding the low temperature CO shift catalyst. For this purpose, the catalyst operates at 200°–250°C which, along with the steam, reduces the ultimate capacity of the ZnO. The zinc oxide is a reagent rather than a catalyst and is discarded after it becomes saturated (7–10% S).

Cobalt oxide–molybdenum oxide on alumina ("cobalt molybdate") or nickel oxide–molybdenum oxide on alumina ("nickel molybdate") type catalysts are coming more and more into favor for removing sulfur. Starting as the oxides, they become converted to the sulfides which can be converted back to the oxides on regeneration. Operating temperatures vary with the service but 320°C is a good average.

The active nickel hydrogenation catalysts have an avid affinity for sulfur and the most active nickel will actually remove sulfur from organic sulfur compounds. These nickel catalysts can be used as "guard cases" to remove sulfur and protect the most sulfur-sensitive catalysts. The nature of the gas stream must obviously be considered for the nickel catalyst may catalyze undesired reactions, e.g.,

$$3\,H_2 + CO \rightarrow CH_4 + H_2O$$

while it is removing sulfur. After becoming sulfided, the nickel "catalyst" is discarded.

A. Regeneration

Iron oxide catalysts are regenerated by steaming, to melt and remove sulfur, or the sulfur is extracted with CS_2 after which the iron sulfides are oxidized by air.

Metallic oxides on activated carbon are regenerated at 150°–175°C with steam at atmospheric pressure and a GHSV of 500 $V_gV_c^{-1}$ hr^{-1}. 800–1500 kilograms of steam are required per cubic meter of catalyst.

The cobalt oxide or nickel oxide–molybdenum oxide on alumina catalysts are regenerated at about 2 atm pressure by burning with air diluted with steam so that the mixture contains about 2.5% O_2. With the GHSV of steam at 800 and of air at 20 $V_gV_c^{-1}$ hr^{-1}, burning begins at about 350°C. The temperature of the "hot spot" as it passes through the bed should not exceed 450°C. After the hot spot has passed through, the steam flow can be reduced to keep the temperature below 475°C. The regeneration is completed when air causes no additional temperature rise.

B. Suppliers of Proven Catalysts

1. *Iron Oxide*

Catalysts and Chemicals, Inc. supplies C9.

Catalysts Division, Chemetron Corp.
Girdler G–42: a promoted iron oxide for use at 250°C.

2. *Metallic Oxides on Activated Carbon.*

Catalysts and Chemicals, Inc.
C8–1: as 4 × 10 and 12 × 30 mesh particles. It contains a lesser content of metallic oxides than C8–6 (below) and is for use with gases containing 8 ppm by volume of organic sulfur compounds or 3 ppm by volume of H_2S maximum.

C8–6: is supplied in the same sizes as C8–1. It can be used with 50 ppm by volume or organic sulfur compounds or 12 ppm by volume H_2S.

Catalysts Division, Chemetron Corp.

Girdler G–32 E: activated carbon + iron oxide as 8 × 30 and 4 × 8 granules.

Girdler G–32 W: activated carbon + copper oxide as 8 × 30 and 4 × 8 mesh granules.

Katalco Corporation

7–1: activated carbon + a promoting metal oxide. The ABD is 0.55, the surface area 1000–1200 m^2gm^{-1}, and the pore volume 0.8 cm^3gm^{-1}.

7–2: activated carbon + a promoting metal oxide. Same properties as 7–1.

3. *Zinc Oxide*

Catalysts and Chemicals, Inc.

C7–1: a 100% ZnO as ¼ × ¼ inch tablets.

C7–2: a zinc oxide–alumina extrudate.

Catalysts Division of Chemetron Corp.

Girdler G–72 C: zinc oxide as ⅛ to 3/16 inch spheres.

Girdler G–72 D: zinc oxide as 3/16 inch extrusions.

Harshaw Chemical Company

Zn–0401 E: 100% ZnO as 3/16 inch extrudates. The ABD is 1.2, the surface area 3 m^2gm^{-1}, and the pore volume 0.26 cm^3gm^{-1}.

Katalco Corporation

32–4: zinc oxide. The ABD is 1.1, the surface area 25 m^2gm^{-1}, and the pore volume 0.30 cm^3gm^{-1}.

4. *Cobalt Oxide and/or Nickel Oxide–Molybdenum Oxide on Alumina Catalysts*

See Chapter 13 for the many suppliers.

5. *Nickel*

See Chapter 13 for the many suppliers.

V. Removing CO, CO_2, or O_2 from H_2. Methanation

Carbon monoxide can be removed from hydrogen streams in two ways. It can be oxidized to carbon dioxide, which is then removed by alkali, or it can be reduced with a part of the hydrogen to form methane. Similarly traces of CO_2 can be hydrogenated to methane without resorting to removal by caustic.

A. Oxidizing CO to CO_2.

When methane from methanation cannot be tolerated, as in metallurgical operations, or is undesirable, as in ammonia synthesis, the CO can be removed by selective oxidation to CO_2 that is scrubbed out by sodium hydroxide (*1*, *3*). Some of the H_2 is oxidized at the same time, so the product gas is usually dried. Ordinarily, the process is applied to gas containing 1% or less of CO. The catalyst is supported platinum (0.1–0.5% Pt), temperature is 50°–150°C, and GHSV 5000–10,000 $V_g V_c^{-1}$ hr^{-1}, with 50–150% excess oxygen. The excess oxygen is needed because the catalyst is not completely selective for CO and some hydrogen is oxidized at the same time. The product gas contains <10 ppm of CO and O_2.

B. Supplier of Proven Catalysts

Engelhard Industries supplies their Selectoxo supported Pt catalysts.

C. Methanation of CO and CO_2 in H_2

Several types of catalysts are available for converting CO and CO_2 to methane. Nickel catalysts are the most common ones. With modern shift catalysts, the gas to be treated will usually contain <1% CO. The operating temperature is 230°–450°C for dry gas and 260°–450°C for gas containing 3–5% H_2O, with GHSV of 1500–5000 $V_g V_c^{-1}$ hr^{-1}, at atmospheric pressure. For 20–30 atm pressure, the GHSV is three to five times that at atmospheric pressure, and for 200–300 atm, the GHSV is ten to fifteen times that at atmospheric pressure.

The reaction is exothermic so there is a temperature rise in the adiabatic catalyst bed. Most catalyst suppliers provide catalysts that can withstand temperatures up to 700°C without serious damage to take care of plant upsets during which excess CO may get into the unit and cause a temperature surge. To use the adiabatic catalyst bed with gases containing more than about 2% CO, a part of the product gas is cooled and recycled to control the temperature rise.

If the hydrogen contains traces of O_2 as well as CO, the nickel catalysts will convert the O_2 to H_2O while converting the CO and CO_2 to CH_4.

Occasionally, some nickel carbonyl will be formed.

It must be kept in mind that these nickel catalysts are highly active for the undesirable reaction:

$$2\,CO = CO_2 + C$$

This reaction can compete with the methanation reaction (*a*) at low H_2:CO ratios and (*b*) at high temperatures. Thus, high catalyst bed temperatures resulting from attempts to methanate more than 2% CO may damage the catalyst both from the temperature and the carbon deposition.

Some of the nickel catalysts are supplied in the oxide form and must be reduced before use. Others are prereduced and are ready for use. The reduction of the oxide catalysts is a strongly exothermic reaction resembling in many respects the regeneration of catalysts by burning off carbon deposits. There is a "hot spot" that passes through the catalyst bed and, unless the maximum temperature of the hot spot is controlled, the catalyst will be damaged. An average reduction schedule is as follows. Heat the catalyst to 250°C in a stream of nitrogen. Gradually bring hydrogen into the circulating nitrogen, keeping the hot spot below 400°C. When pure hydrogen can be used, raise the temperature of the whole catalyst bed to 400°C. The catalyst is now ready to use. The reduced catalyst is pyrophoric, so adequate precautions must be taken if the catalyst is reduced in one vessel and used in another.

Whether the original catalyst was supplied as oxide or prereduced, the used catalyst is pyrophoric and must be stabilized before exposing it to the air. Each catalyst supplier will provide an appropriate stabilization procedure. The following is an average one. Cool the catalyst to 100°C in circulating nitrogen. Bleed air into the nitrogen at a rate that keeps the hot spot temperature below 250°C. When the hot spot has passed through the bed, cool the catalyst and it can be handled in air. If the catalyst is not spent, it is in a prereduced condition and stabilized, ready for reuse.

D. Poisons

All the nickel catalysts are permanently poisoned by sulfur and arsenic compounds. Especial care is needed, if the CO_2 scrubber earlier in the hydrogen preparation system contains arsenic or sulfur compounds, to prevent carry over into the nickel catalyst. With reasonable care, the catalyst life is 2–3 years. After amine or carbonate removal systems, the life should be >5 years.

In addition to the nickel catalysts, Pt and Ru catalysts are also used for methanation. These catalysts do not have to be reduced or stabilized before use and do not become pyrophoric during use, so they do not have to be stabilized before removing from the reactor. They are particularly attractive for installations where the cost of reduction or stabilization equipment cannot be justified.

E. Suppliers of Proven Catalysts

Catalysts and Chemical Inc.

C13–3: nickel oxide on alumina as $\frac{3}{8} \times \frac{3}{16}$ inch tablets.

C13–4: nickel oxide on alumina as $\frac{3}{16}$–$\frac{5}{16}$ inch spheres. The C13–4 is more active than the C13–3.

C53: a supported Pt–Ni.

Catalysts Division of Chemetron Corp.

G–33: 33% Ni on a refractory support as ¼ inch tablets. The ABD is 1.

G–52: the reduced and stabilized form of G–33.

G–65: 25% Ni on a support, as tablets; in oxide or prereduced and stabilized form.

G–65S: 17% Ni on a support, as spheres; in oxide or prereduced and stabilized form.

Engelhard Industries supplies their Type E, supported Ru.

Harshaw Chemical Company

Ni–0101 T: an unreduced nickel hydrate on kieselguhr, containing 44% Ni, as ⅛, $\frac{3}{16}$, or ¼ inch tablets. The ABD is 1.2–1.35.

Ni–0104 T: a reduced and stabilized catalyst containing 58% Ni on kieselguhr, as ⅛, $\frac{3}{16}$, or ¼ inch tablets. The ABD is 1.35–1.55, the surface area 125 m^2gm^{-1}, and the pore volume 0.2 cm^3gm^{-1}.

Katalco Corporation supplies:

11–3: nickel oxide on alumina. The ABD is 1.0, the surface area 60–80 m^2gm^{-1}, and the pore volume 0.2–0.3 cm^3gm^{-1}.

Matthey Bishop, Inc. supplies their THT supported Pt type catalysts.

VI. Removing O_2 or H_2

The familiar laboratory technique of using reduced copper oxide to remove O_2 is also used industrially, as is the technique of using copper oxide to remove H_2. These are reagents rather than catalysts.

For removing O_2, the Cu must be active and is preferably prepared by reducing the oxide at 175°–225°C with hydrogen. The reduction is not carried all the way to red copper but is stopped just as the black begins to turn red. It is used at 250°–450°C.

When the O_2 is in H_2, they can be made to combine by a nickel catalyst at 200°–400°C, and this is a true catalyst rather than a reagent.

For removing small amounts of O_2 from H_2 or H_2 from O_2, several platinum group catalysts are available. Supported Pd is effective at temperatures from below 0°C to 1000°C. CO acts as an inhibitor so that certain minimum temperatures are necessary, e.g., with 10 ppm CO, 20°C or 1000 ppm CO, 70°C. Supported Pt and Ru catalysts are effective at about the same conditions as Pd. These catalysts are also effective for combining the H_2 and O_2

or D_2 and O_2 formed in water or deuterium oxide moderated nuclear reactors.

A. Poisons

The copper reagents are subject to sulfur poisoning which prevents regeneration. The nickel catalysts are poisoned by sulfur and arsenic. The Pt group metals are poisoned by chlorides, sulfur, arsenic, and lead and inhibited by CO, olefins, and lubricating oils.

B. Suppliers of Proven Catalysts

Catalysts and Chemicals, Inc.
C53: a Pt–Ni on alumina.
C54: a series of Pd on activated alumina catalysts.

The Air Correction Division of Universal Oil Products Co. supplies standard elements of Pt supported on a high nickel alloy ribbon that is fabricated into metal screens and mats. These are mounted in complete reactor units.

Engelhard Industries supplies a copper type reagent "C," as well as their supported Pd catalyst "D" in the form of pellets.

Harshaw Chemical Company
Cu–0307 T: 99% CuO as ⅛ inch tablets. The ABD is 1.35, the surface area 8 m^2gm^{-1}, and the pore volume 0.26 cm^3gm^{-1}.
Cu–0803 T: 10% CuO on high activity alumina as ⅛ inch tablets. The ABD is 1, the surface area 137 m^2gm^{-1}, and the pore volume 0.42 cm^3gm^{-1}.
Cu–1700 T: 79% CuO as ⅛ inch tablets. The ABD is 1.7, the surface area 8 m^2gm^{-1}, and the pore volume 0.14 cm^3gm^{-1}.
Cu–1900 S: 5% CuO on inert alumina as ½ inch spheres. The ABD is 1.2, the surface area 1 m^2gm^{-1}, and the pore volume 0.07 cm^3gm^{-1}.
Ni–0104 T: prereduced and stabilized 58% Ni on kieselguhr as ⅛, 3/16, or ¼ inch tablets. The ABD is 1.35–1.55, the surface area 125 m^2gm^{-1}, and the pore volume 0.2 cm^3gm^{-1}. Although prereduced, the catalyst should be given a short H_2 treatment at 400°C before this service.

Matthey Bishop, Inc. supplies supported Pd, Pt and Ru catalysts.

VII. Removing Nitrogen Oxides from H_2, CO, Natural Gas, or the Exhaust from a Nitric Acid Plant

Several chemical operations produce nitrogen oxides, e.g., the oxidation of ammonia to make nitric acid, the use of nitric acid as an oxidizing agent, etc. These gases must be made tolerable to the surrounding community before venting to the atmosphere. This is done by causing them to react with a reducing agent, H_2, CH_4, heavier hydrocarbons, coke oven gas, etc., in the presence of a supported Pd or Pt catalyst. The reaction is exothermic, so the temperature rise in an adiabatic reactor depends on the total oxygen (including that in the nitrogen oxides) in the initial gas. The rise is 130°–140°C with CH_4 as the fuel (reducing agent) or 150°–160°C with H_2 as the fuel for each percent oxygen in the inlet gas.

The minimum inlet temperature is about 150°C with H_2 as the fuel, 350°C with naphtha using a GHSV of up to 100,000 $V_g V_c^{-1}$ hr^{-1}. With natural gas as the fuel, the inlet temperature should be 450°C or higher at a GHSV of up to 30,000 $V_g V_c^{-1}$ hr^{-1}. With these high space velocities, the catalysts and catalyst beds are designed for low pressure drop, as low as 100 mm Hg. The total pressure can vary from substantially atmospheric to as high as the rest of the process might demand. For example, in the ammonia oxidation tail gas operation, 5–7 atm is common.

The reaction is exothermic so there is a temperature rise in an adiabatic bed. For several reasons, the maximum temperature should be kept below about 800°C. This means that, with H_2 as the fuel, the maximum permissable oxygen is about 4% and, with natural gas, about 2.5%. The fuel requirement is slightly in excess of the stoichemetric needs of the total oxygen present. Nitrogen oxides in the treated gas are 200 ppm or less.

Nitrogen oxides are present in the exhaust from automobiles and are objectionable because they take part in the formation of smog. As of this writing, a proven catalyst for reducing these oxides to an acceptable level when the engine is using fuels containing lead alkyls is not available.

A. Suppliers of Proven Catalysts

Catalysts and Chemicals, Inc. supplies C53, a Pt–Ni on alumina catalyst, as well as the C54 series of Pd on activated alumina catalysts.

Catalysts Division, Chemetron Corp.
Girdler G–43: Pt–Ni on alumina as ¼ and ⅜ inch tablets. The ABD is 1.

Engelhard Industries supplies HCD for easily oxidized fuels and D (¼ inch spheres) for natural gas. Both are supported Pd catalysts.

Matthey Bishop, Inc. supplies their THT Pt type metals on honeycomb ceramic supports.

VIII. Removing Diolefins and Acetylenes from Olefin Streams

It is common practice to produce ethylene and propylene for petrochemical use by the high temperature steam cracking of propane, naphtha, or other petroleum fractions. Whenever this is done, the raw gas contains a certain amount of acetylene, methylacetylene, propadiene, and residual butadiene. These are ordinarily not wanted in an olefins petrochemical operation, and it is customary to hydrogenate them to the corresponding monoolefins and so add to the total yields of the products. This hydrogenation may be carried out on the raw gas or on the partly or almost completely purified (concentrated olefin) gas. Since the catalysts differ for these two operations, they will be discussed separately. In general, fairly rugged catalysts are employed on the raw gas and these operate under moderately severe conditions, while more selective catalysts are used with mild conditions on the concentrated olefins. In either type of operation, there is some polymer formed on the catalyst which reduces its effectiveness. The catalysts must be regenerated every 3–6 months with steam and air or nitrogen and air to remove the polymer. After five to ten regenerations the catalyst is replaced. Modest temperature increases (20°–30°C) during operation are used to compensate for the loss in activity due to polymer deposition. In many plants it is more economical to replace the polymer coated catalyst with fresh catalyst than to absorb the cost of the nonproductive time required to regenerate.

A. Raw Gas

There are certain advantages and disadvantages to removing the acetylenes and dienes from the raw gas. The needed hydrogen is already present. This is not an unmixed blessing, because it can cause a "runaway" hydrogenation. The selective hydrogenation is exothermic, and if the temperature is not carefully controlled in the adiabatic fixed bed reactor, the hydrogen can start reacting with the monoolefins, generating more heat and causing the runaway.

If there is butadiene in the raw gas, it is also hydrogenated, so this operation is not used where butadiene is to be recovered (*2*).

Supported Pd catalysts are used under the following conditions: Temperature 35°–100°C, 3–30 atm pressure, and GHSV = 4000–7000 $V_g V_c^{-1}$hr^{-1}. Nickel sulfide or nickel–cobalt–chromium catalysts are used as follows: temperature 120°–300°C, pressure 3–30 atm, and GHSV = 1000–3000 $V_g V_c^{-1}$ hr^{-1}. These catalysts operate in the presence of sulfur compounds (usually present) which help prevent runaway hydrogenation.

B. Ethylene Streams

On concentrating the ethylene, the hydrogen originally present was removed, so hydrogen must be added to react with the C_2H_2. The amount is

2 $H_2:C_2H_2$–3 $H_2:C_2H_2$. To operate near stoichiometric, i.e., 1.2 $H_2:C_2H_2$, the space velocity must be reduced drastically, e.g., to one-tenth the velocity used with excess hydrogen.

The most widely used catalysts are low Pd (<0.05%) on alumina supports. Operating conditions are within the following ranges: temperature 40°–150°C, pressure 1–35 atm, and GHSV = 2000–4000 $V_g V_c^{-1}$ hr^{-1} for $H_2:C_2H_2$ of 2:1–3:1. These conditions will vary with the impurity level. Also the presence of sulfur compounds can cause some inhibition and require conditions in the more severe part of the range.

C. Propylene Streams

Both propadiene and methylacetylene are found in pyrolysis propylene. They are selectively hydrogenated by Pd on alumina or Pd–chromia catalysts. Operating conditions (*2*) are $H_2:C_3H_4$ 3:1, temperature 150°–200°C, pressure 1–20 atm, and GHSV = 1300–4500 $V_g V_c^{-1}$ hr^{-1}, depending on the impurity level.

D. Butylene Streams

Cobalt oxide–molybdenum oxide–alumina (*6*), cobalt oxide–nickel oxide–molybdenum oxide–alumina, and nickel oxide–molybdenum oxide–alumina catalysts in sulfided form have been considered for selectively hydrogenating butadiene in butylene streams. (Chapter 9 gives an illustration of the deleterious effect of butadiene in butylene in isobutane alkylation.) Laboratory conditions with sulfided cobalt oxide–molybdenum oxide–alumina catalyst were 260°C and atmospheric pressure. A small amount of CS_2 (0.02% of feed) had to be added to the feed to maintain catalyst activity. The writer is not familiar with a commercial operation using these catalysts.

E. Butadiene Streams

A proprietary catalyst is available for hydrogenating acetylenes in streams containing up to 80% butadiene. The acetylenes are reduced to <50 ppm.

F. Removing Excess Hydrogen

When all the impurity has been hydrogenated, a small amount of excess hydrogen may remain. This causes no difficulty in many operations but its removal is required in some. A higher Pd content (about 0.5%) on alumina catalyst will cause the hydrogen to react with the olefin present at temperatures in the 100°C–200°C range. The catalyst is often placed in the exit portion of the main catalyst bed.

G. Suppliers of Proven Catalysts

Catalysts and Chemicals Inc.

C31–1: a low Pd on a high surface area alumina primarily for propylene rich streams; as 4–8 mesh spheres.

C31–1A: a low Pd on a lower surface area alumina for ethylene rich streams; as 4–8 mesh spheres.

C31–3: a higher Pd content on alumina catalyst for causing excess H_2 to react with the olefin present.

C35–2: a Pd on alumina for treating hydrogen containing raw pyrolysis gas, especially where the C_4 hydrocarbon content is low; as ⅛ inch extrudates.

C36–1, C36–2, and C36–3: nickel, cobalt, chromium on alumina for treating raw pyrolysis gas containing hydrogen in the presence of sulfur compounds; as ⅓ inch spheres. The activity of the catalysts decreases in the order C36–1 > C36–2 > C36–3.

Catalysts Division of Chemetron Corp.

G–55B: a Pd–chromia catalyst for propylene rich streams.

G–58B: a supported Pd catalyst for ethylene rich streams.

G–73: an inhibited nickel catalyst for raw gas streams.

G–83: a Pd on alumina catalyst for raw gas streams.

Dow Chemical Company provides their Type P hydrogenation catalyst (proprietary) for ethylene rich streams and Type K for acetylenes in butadiene rich streams (also proprietary).

Engelhard Industries, Inc. supplies a series of low Pd catalysts K–0, K–1, K–2, KM, KL, and KH.

Harshaw Chemical Company supplies Ni–0301 T, a nickel oxide (11% Ni) on activated alumina as ⅛ inch tablets. ABD is 1.1, surface area 64 m^2gm^{-1}, and pore volume 0.32 cm^3 gm^{-1}. The catalyst may be sulfided before use as follows: heat the catalyst to 310–315°C and, at atmospheric pressure, pass 10% H_2S in H_2 over the catalyst at GHSV 600 $V_gV_c^{-1}$ hr^{-1}. After the hot spot has passed through the bed, keep the gas mixture flowing for 30 minutes and then raise the temperature to 370°C over a period of one hour. Continue the gas flow at 370°C for two hours and cool to 235°C in the gas. Raise the pressure to operating pressure with H_2, and the temperature to operating temperature.

Katalco Corporation supplies 38–1, Pd on alumina. ABD is 1.05, surface area 20–30 m^2gm^{-1}, and pore volume 0.2–0.3 cm^3 gm^{-1}.

For the many suppliers of the cobalt oxide–molybdenum oxide–alumina type catalysts see Chapter 16.

REFERENCES

1. Andersen, H. C., and Green, W. J., *Ind. Eng. Chem.* **53,** 645 (1961).
2. Andersen, H. C., Haley, A. J., and Egbert, W., *Ind. Eng. Chem.* **52,** 901 (1960).
3. Brown, M. L., Jr., Green, A. W., Cohn, G., and Andersen, H. C., *Ind. Eng. Chem.* **52,** 841 (1960).
4. Christian, J. G., and Johnson, J. E., *Am. Chem. Soc. Div. Petrol. Chem. Preprints* **7,** No. 3, 63 (1962); *Air Water Pollution* **9** (1/2), 1 (1965).
5. Johnson, J. E., Christian, J. G., and Carhart, H. W., *Ind. Eng. Chem.* **53,** 900 (1961).
6. Kirsch, F. W., and Shull, J. E., *Ind. Eng. Chem. Prod. Res. Develop.* **2,** 48 (1963).
7. Kohl, A. L., and Riesenfield, F. C., "Gas Purification." McGraw-Hill, New York, 1960.
8. Rushton, J. H., and Krieger, K. A., *Advan. Catalysis* **3,** 107 (1951).
9. Yarrington, R. M., and Bambrick, W. E., personal communication, 1969.

CHAPTER 13

HYDROGENATION

There are more hydrogenation catalysts available commercially than any other type. As a result, the reader seeking a catalyst is more likely to be confused by the wide choice than by other factors. In an attempt to help the reader, many of the commercial hydrogenation processes have been described in sections or chapters elsewhere in this book. The present chapter is intended to be illustrative rather than complete. It shows how the catalysts are used in hydrogenating relatively simple organic compounds. The reader is doubtless aware that whole books have been written on hydrogenation and hydrogenation processes (*1*, *4*, *8*, *9*, *21*, *22*, *26–28*). Advantage has been taken of this by severely limiting the references in this chapter to those that serve the purpose of giving the reader some idea of operating conditions under which the catalysts are used.

I. Ortho-Para Hydrogen Conversion

Although not a true hydrogenation, this intriguing process is appropriately treated here because its study has contributed so much toward a partial understanding of the catalytic hydrogenation process (*4*). The more practical aspects of the reaction are considered here as they affect the liquefication and storage of liquid hydrogen. At liquid hydrogen temperatures, the equilibrium concentration of p-H_2 is 100%. However, at room temperature it is 25% the conversion of the *ortho* to the *para* is exothermic. Freshly liquefied hydrogen may contain close to 75% *ortho* which slowly converts to *para* liberating the heat of reaction and boiling away hydrogen. After 100 hours of storage, 40% of the original liquid evaporates due to this heat release. It is therefore desirable to effect this conversion during the liquefication process and liberate the heat at as high a temperature as practical (*25*). To do this, hydrogen streams are diverted through fixed catalyst beds at several decreasing temperatures during the liquefication. The final liquid hydrogen stored is essentially 100% *para*. Hydrous ferrous oxide is one of the more active oxide catalysts (*25*) and supported Ru is a very active metal catalyst (*5*).

When liquid hydrogen is to be used at some higher temperature than the liquid, and the refrigeration provided by the boiling liquid is needed, additional refrigeration can be obtained from the *para–ortho* conversion, which is endothermic. The same catalysts may be used (*5*).

A. Suppliers of Proven Catalysts

Both Engelhard Industries and Johnson Matthey, Inc. supply supported Ru catalysts.

The writer does not know of a commercial supplier of the hydrous ferrous oxide gel catalyst.

II. Hydrogenation of Olefins

A. Nickel Catalysts

If they are free of catalyst poisons, olefins can be hydrogenated at atmospheric pressure and room temperature with active nickel catalysts. Ordinarily, however, both temperature and pressure are raised. The conditions used in a former process for hydrogenating diisobutylene and a codimer of iso- and *n*-butylene illustrate commercial practice. A fixed bed of tablets of about 60% reduced nickel on kieselguhr was used at 4 atm pressure and 175°–200°C. The reaction is strongly exothermic so that the elevated temperature is as much a convenience for heat removal as for controlling the rate of reaction.

These catalysts are just as useful in stirred reactors. The catalyst is a powder often supported on a filter aide (kieselguhr) to help remove the catalyst at the end of the reaction (the powdered catalysts become colloidal and difficult to remove). Reaction temperatures are usually in the range of 50°–150°C, with pressures of 5–10 atm. The amount of catalyst is 2–5% by weight. Occasionally, the initial addition of catalyst is poisoned by unidentified material in the charge. If the amount of poison is not too great, a second addition of catalyst will cause hydrogenation.

The nickel catalysts are sensitive to CO in the H_2 feed as well as sulfur or halogen compounds, arsenic and lead.

Because of the many different nickel catalysts offered commercially, it is best to divide them into several categories, discuss each briefly, and list the suppliers of each type. The types are:

(*a*) Supported nickel oxide, hydrate, or carbonate, all of which must be decomposed and reduced before use.

(*b*) Reduced and stabilized nickel catalysts that are ready to be used.

(*c*) Reduced and active nickel catalysts that are normally pyrophoric but are supplied in a protective medium.

(*d*) Raney alloy, a Ni–Al alloy that is activated by aqueous alkali before use.

Because these catalysts can be used for hydrogenating many substances other than olefins, the following discussion is broadly applicable.

B. Supported Nickel Oxide, Hydrate, or Carbonate

Most of these catalysts are in the form of tablets or extrudates and are used in fixed bed adiabatic reactors. Although the nickel compound, the nickel content, and the support all vary widely, the following activation procedures are usable with all of them. Assuming that the catalyst has been loaded into an adiabatic reactor, heat to 385°C by passing hot nitrogen over the catalyst to decompose the hydrate or carbonate and then cool to 335°C in the nitrogen. Bleed hydrogen into the nitrogen while keeping the maximum temperature below 385°C. Depending on the nickel content, there will be a more or less well-defined hot spot that will pass through the bed. Keep raising the hydrogen content of the circulating gas stream until pure hydrogen can be used without exceeding 385°C. The catalyst is cooled to hydrogenation temperature and is ready to use. It is pyrophoric and should not be exposed to air.

An alternate procedure that is more convenient in some cases is as follows. Purge the catalyst bed of air (ambient temperature) with nitrogen and then replace the nitrogen with hydrogen. Increase the temperature of the hydrogen slowly, no more than 90°C per hour, until a temperature of 400°C is reached. Hold at 400°C for 5 hours and then cool to reaction temperature. The catalyst is pyrophoric and should not be exposed to air.

1. *Suppliers of Proven Catalysts*

Catalysts Division of Chemetron Corp.

G–33: 33% Ni on a refractory oxide support as ¼ inch tablets. The ABD is 1.

Harshaw Chemical Company

Ni–0101T: 44% Ni (as hydrate) on kieselguhr; ⅛, 3/16, and ¼ inch tablets. The ABD is 1.2–1.4.

Ni–0102T: 55% Ni (as oxide) on kieselguhr; ⅛, 3/16, and ¼ inch tablets. The ABD is 1.5–1.7.

Ni–0108T: 66% nickel oxide on kieselguhr; ¼ inch tablets. The ABD is 1.2.

Ni–0302T: 15% Ni (as oxide) on activated alumina; ⅛ inch tablets. The ABD is 1.3, the surface area 65 m^2gm^{-1}, and the pore volume 0.20 cm^3gm^{-1}.

Ni–0304T: 32% nickel oxide on active alumina; 5/32 inch tablets. The ABD is 1.4, the surface area 52 m^2gm^{-1}, and the pore volume 0.23 cm^3gm^{-1}.

Ni–0704E: 18% Ni (as oxide) on kieselguhr; 5/32 inch extrudates. The ABD is 0.85.

Ni–0707T: 14% nickel on highly active alumina; ⅛ inch tablets. The ABD is 0.85, the surface area 140 m^2gm^{-1}, and the pore volume 0.38 cm^3gm^{-1}.

Ni–1406T: 46% Ni on a special support; 5/32 and 7/32 inch tablets. The ABD is 1.3–1.4.

Ni–3207T: 28% Ni on a special support; ⅛, 3/16, and ¼ inch tablets. The ABD is 1.1–1.2.

C. Reduced and Stabilized Catalysts

These catalysts are similar to the ones described above except the catalyst supplier has already reduced the catalyst. The reduction is rarely complete and in fact the preferred catalysts usually contain some unreduced oxide. As produced, these catalysts are pyrophoric. To make them safe to handle in air, they are given a mild, partial oxidation treatment treatment. The catalysts are ready to use as received. Some of the catalysts are supplied as powders to be used suspended in the liquid or solution being hydrogenated.

1. *Suppliers of Proven Catalysts*

Catalysts Division of Chemetron Corp.

G–49: 65% Ni supported on kieselguhr; as a powder with 90% passing a 325 mesh screen. The ABD is 0.4.

G–52: the reduced and stabilized form of G–33, above (Section II B1).

G–69: a zirconium promoted catalyst containing 50% Ni and 2% Zr. The ABD is 0.5 and 90% passes a 325 mesh screen.

Harshaw Chemical Company supplies:

Ni–0104P: 58% Ni on kieselguhr in which 60–65% of the total nickel has been reduced. The powder ABD is 1.1 and the surface area 155–165 m^2gm^{-1}.

Ni–0104 T: a tabletted form of Ni–0104 P; 1/8, 3/16, and 1/4 inch tablets. The ABD is 1.4–1.5, surface area 125 m^2gm^{-1}, and the pore volume 0.2 cm^3gm^{-1}.

Ni–0107 T: 56% Ni on kieselguhr in which about one-half the nickel has been reduced. The ABD is 1.5, the surface area 130 m^2gm^{-1}, and the pore volume 0.18 cm^3gm^{-1}.

Ni–0116 T: 60% on kieselguhr as 1/8, 3/16, or 1/4 inch tablets. The ABD is 1.4–1.6, the surface area 155–165 m^2gm^{-1}, and the pore volume 0.22 cm^3gm^{-1}.

Ni–0122 E: 28% Ni on kieselguhr; 1/16 inch extrudates. The ABD is 0.75, the surface area 168 m^2gm^{-1}, and the pore volume 0.53 cm^3gm^{-1}.

Ni–1404 P Spec. 103: 68% Ni on a special support; a powder with an ABD of 1.7 and a surface area of 115 m^2gm^{-1}.

Ni–1414 T, Spec. 103: the tabletted form of Ni–1404 P; 1/8 and 3/16 inch tablets. The ABD is 1.0–1.1, the surface area 125 m^2gm^{-1}, and the pore volume 0.35 cm^3gm^{-1}.

Ni–1430 E: 40% Ni on alumina; 1/16 inch extrudates. The ABD is 0.65, the surface area 160 m^2gm^{-1}, and the pore volume 0.69 cm^3gm^{-1}.

Ni–3210 T: 35% Ni on a special support for trickle beds or aqueous substrates; 1/8, 3/16, and 1/4 inch tablets. The ABD is 0.8–0.9, the surface area 175 m^2gm^{-1}, and the pore volume 0.45 cm^3gm^{-1}.

Universal Oil Products Co.

H–2: 60% Ni on kieselguhr as 1/8 inch extrudates. The ABD is 1.1.

H–4: 30% Ni on alumina as 1/8 inch spheres. The ABD is 0.6–0.65.

D. Reduced and Active (Pyrophoric) Catalysts

These catalysts have been prepared in highly active form by the catalyst manufacturer. An original oxide may have been carefully reduced or a Raney alloy (Ni–Al) may have been caustic leached to produce the active catalyst. The catalysts are pyrophoric so they are shipped in a protective liquid (e.g., water) or solid (e.g., stearin). They are most commonly used as powder slurrys in the liquid to be hydrogenated in a stirred reactor that can be heated to start the reaction and later cooled to remove the heat of hydrogenation. During use, the catalysts become very finely divided and difficult to remove by filtration, so the catalyst is sometimes supplied on a carrier or mixed with a filter aide. The catalysts are most often used for the hydrogenation of fats and oils (Section V) but are also active for olefinic hydrocarbons.

1. *Suppliers of Proven Catalysts*

Catalysts supplied specifically for hydrogenating fats and oils are listed in that section.

Davison Chemical Division of W. R. Grace & Co.

Grade 24: active chromium promoted Ni from Raney alloy as a finely divided powder in water.

Grade 28: active Ni from Raney alloy as a finely divided powder in water.

Grade 4200: active Ni from Raney alloy as a coarse powder in water.

Harshaw Chemical Co.

Ni–0109 T: 60% Ni on kieselguhr in which 75% of the Ni has been reduced; ¼ inch tablets. The ABD is 1.3 and the catalyst is shipped in a protective liquid.

E. Raney Nickel Alloys

Ni–Al alloys (*17*) are available for those who want to activate them with sodium hydroxide and use them immediately or soon after preparation for maximum utility of the hydrogen adsorbed on the Ni. The alloys are supplied as powders or as granules. The powders are used suspended in the liquid in stirred reactors. The granules are put in a fixed bed and surface activated where they are most often used with vapors plus hydrogen or in trickle bed operations.

The activity of the catalyst depends to some extent on the method of caustic treating the alloy. The following procedures may be useful. Prepare a 25% NaOH solution in which the weight of NaOH (anhydrous) is equal to three-quarters the weight of 50:50 Ni–Al alloy. Add the alloy powder to the solution as fast as possible without uncontrolable foaming. Digest the mixture for 3 hours near the boiling point, adding water to maintain a constant volume. Wash free of alkali, protecting the catalyst from the air at all times, i.e., do not permit the solid to suck dry on a filter. The dry catalyst is pyrophoric and even the moist catalyst will react slowly with oxygen. Tap water is preferred to distilled or demineralized water.

Hydrogen is liberated during the caustic digestion and reaction, so use adequate fire precautions. The catalyst still contains some aluminum as well as adsorbed hydrogen. It slowly evolves hydrogen, so it must not be stored under water in a closed container.

The water may be displaced by higher boiling liquids simply by adding the liquid and heating or by simple displacement with water-soluble liquids. Exposure to air during displacement must be avoided.

The granular catalysts in the fixed bed are activated by circulating a 10% sodium hydroxide solution through the bed until about 10% of the aluminum has been removed. The catalyst is washed free of alkali and is ready for use. The catalyst must not be exposed to air in this operation.

1. *Suppliers of Proven Catalysts*

Davison Chemical Division of W. R. Grace & Co.
Grade 2428: a chromium promoted nickel aluminum alloy powder.
Grade 2813: a 50% Ni–50% Al alloy powder.
Grade 5842: a Ni–Al alloy as granules.

Harshaw Chemical Company
Ni–1000P: a 50% Ni–50% Al alloy powder, 84% passing 325 mesh.
Ni–3001T: a 49% Ni–49% Al alloy in the form of 3/16 inch pellets.

F. Used Nickel Catalysts—Safety Precautions

Almost all nickel catalysts become pyrophoric when used in hydrogenations. For safety it must be assumed that a used catalyst is pyrophoric unless it is known for certain that it is not. For protection, a used catalyst may be "drowned" in water or other liquid that will prevent exposure to air. The mixture may then be handled in air as long as it is protected by the liquid.

The fixed bed supported catalysts can be stabilized just as the freshly reduced catalysts are stabilized. The following stabilization procedure may be useful. The author is not sure the procedure will work for beds of surface activated Raney alloy catalysts.

Purge the catalyst with nitrogen and cool to 100°C. Bleed air into the circulating nitrogen, keeping the peak temperature below 250°C. When the hot spot has passed through the bed, cool to ambient temperature and the catalyst can be handled in air. If the catalyst is not spent, it can be reloaded and reused.

Some used catalysts have organic materials adhering or deposited on them, so the user should be alert for a true burning wave going through the catalyst bed. If there is one, it will be virtually impossible to keep the temperature below 250°C and even below 400°–500°C. Such a burning wave oxidizes the nickel so that it is not active without reduction and even then its activity is usually impaired by the high temperature.

G. Platinum Metals

An entire book is devoted to hydrogenations with platinum type metals (*22*), so it is deemed adequate here to list the suppliers of proven catalysts. Frequently, one of these catalysts will be an active hydrogenation catalyst for double bonds, even though an element is in the molecule to be hydrogenated that would poison nickel catalysts, e.g., chloro or bromo organics. The catalysts are widely used in the chemical and pharmaceutical industries in proprietary operations.

Engelhard Industries, Inc. or Johnson Matthey, Inc. supply a wide variety of these catalysts, such as the ones listed in Table VII. The common metal contents are 0.5% and 5.0%.

TABLE VII

PROVEN PLATINUM TYPE CATALYSTS SUPPLIED BY ENGELHARD INDUSTRIES, INC. OR JOHNSON MATTHEY, INC.

Metal	Support	Form
Pd	Charcoal	Powder or paste (in water)
Pd	Charcoal	Granules or extrudates
Pt	Charcoal	Powder or paste (in water)
Pt	Charcoal	Granules or extrudates
Ru	Charcoal	Powder or paste (in water)
Ru	Charcoal	Granules or extrudates
Pd	Alumina	Powder, granules, tablets, or extrudates
Pt	Alumina	Powder, granules, tablets, or extrudates
Ru	Alumina	Powder, granules, tablets, or extrudates
Pd	$CaCO_3$	Powder
Rh	Charcoal	Powder, granules, or extrudates
Rh	Alumina	Powder, granules, tablets, or extrudates

Calsicat Division of Mallinckrodt Chemical Works

Pd on carbon.

Pt on carbon.

Catalysts Division of Chemetron Corp.

G–71 and G–81: Pd on powdered carbon, dry or paste (in water). The following letter suffixes indicate the Pd content. The ABD is 0.4.

A	1%
B	3%
C	5%
D	10%

Davison Chemical Division of W. R. Grace & Co.

Grade 936: 5% Pd on charcoal supplied as a wet paste (55% water) containing 0.6% Na_2O.

Grade 937: 5% Pd on charcoal; like 936 except 0.06% Na_2O.

Both of these catalysts are pyrophoric in the dry state.
Grade 942: 0.2% Pd on 4–6 mesh carbon granules.

Harshaw Chemical Company
Pd–0501 T: 0.3% Pd on high activity alumina, as ⅛ inch tablets. The ABD is 0.9, the surface area 186 m^2gm^{-1}, and the pore volume 0.40 cm^2gm^{-1}.
Pd–0505 T: 0.3% Pd on high activity alumina, activated and stabilized, as ⅛ inch tablets. The ABD is 0.9, the surface area 200 m^2gm^{-1}, and the pore volume 0.40 cm^3gm^{-1}.
Pd–0509 T: 0.3% Pd (activated) on a special support, as ⅛ inch tablets. The ABD is 1.1, the surface area 180 m^2gm^{-1}, and the pore volume 0.40 cm^3gm^{-1}.

Universal Oil Products Co.
H–3: 0.75% Pt on alumina as ⅛ inch spheres.

H. Noble Metal Recovery

Suppliers of noble metal catalysts have arrangements for recovering the metal from spent commercial catalysts for reuse. It is this fact that permits the noble metals to compete where otherwise their cost would be prohibitive. Base metal catalysts are usually discarded after being spent.

I. Sulfur-Resistant Catalysts

It is not always practical to remove enough sulfur from olefins to permit hydrogenation with the sulfur-sensitive catalysts. There is a large number of so called "cobalt molybdate" type catalysts available for this service. These are cobalt oxide–molybdenum oxide–aluminum oxide, or the same catalyst in which part or all of the cobalt oxide has been replaced with nickel. Sulfur combines with the original catalyst to form the active species. Unless there is sulfur in the feed, the catalyst loses activity. While not essential with the cobalt oxide–molybdenum oxide catalysts, it is best to presulfide the nickel oxide–molybdenum oxide catalysts.

Operating conditions for hydrogenating a mixture of octenes are (*15*): Temperature 235°–340°C, 25–35 atm pressure, LHSV 1–2 $V_1V_c^{-1}$ hr^{-1}, with hydrogen feed of 2–4 moles per mole of olefin.

1. *Suppliers of Proven Catalysts*

See Chapter 16 for suppliers and catalyst sulfiding procedure.

III. Hydrogenation of Acetylenes to Olefins

Chapter 12 gives information on the selective hydrogenation of small concentrations of acetylenes in olefin streams. There are processes in which the entire substance to be hydrogenated is an acetylene. For example, a process has been developed for making isoprene in which acetylene is condensed with acetone to produce 3-methylbutynol-3. This is selectively hydrogenated to 3-methylbutenol-3, which is catalytically dehydrated to isoprene. The hydrogenation conditions are 30°–80°C and 5–10 atm pressure, using a continuous reactor containing slurried, supported catalyst. The catalyst is inhibited Pd (*6*). The preferred catalyst for this type of selective hydrogenation is Pd supported on $CaCO_3$, inhibited with lead acetate and a little quinoline (*18*, *22*).

A. Suppliers of Proven Catalysts

Both Engelhard Industries and Johnson Matthey, Inc. supply Pd on $CaCO_3$.

IV. Steam Cracker Naphtha

In the high temperature steam cracking of naphthas to produce olefinic gases for chemical syntheses, a highly aromatic, highly unsaturated residue is left. Most of this residue is within the motor fuel boiling range and it has excellent antiknock quality. However, the unsaturated hydrocarbons polymerize during distribution and these polymers deposit in tanks and pipelines. The objective is to hydrogenate these unsaturates selectively, without hydrogenating the aromatic hydrocarbons and without damaging the antiknock quality. The unsaturated materials are hydrocarbons such as cyclopentadiene and its dimer, styrene, indene, etc.

Supported nickel catalysts are used in some of the processes. The sulfur in the raw material does act as an inhibitor, in that it slows down the reaction, but at the same time, it makes the catalyst more selective.

In one process (*2*, *16*, *29*), the operating conditions are: temperature 125°–180°C, 20–40 atm pressure, LHSV 0.5–5.0 $V_l V_c^{-1}$ hr^{-1}, with 0.5 moles of H_2 per mole of total hydrocarbon.

In a second process the conditions are a little more severe (*14*, *20*): temperature 180°C, 50 atm pressure, LHSV 2 $V_l V_c^{-1}$ hr^{-1}, with 0.5–3 moles of H_2 per mole of total hydrocarbon.

As might be expected, there is a tendency for some of the unsaturated hydrocarbons to polymerize on the catalyst. Evidently, keeping the temperature low enough, the pressure high enough, and the H_2:hydrocarbon ratio low enough to have some liquid phase present retards the polymer formation on the catalyst (*14*, *20*). Ultimately, the catalyst must be regenerated, using air plus an inert gas, but keeping the maximum temperature below 550°C.

Pd catalysts are also used in this process at 25°–200°C, 10–70 atm pressure, and LHSV 1–6 $V_l V_c^{-1}$ hr^{-1}, with hydrogen additions depending on the diolefin content. Limited amounts of sulfur compounds are tolerated, e.g., 200 ppm.

Extensive pilot plant work has been done using a commercial supported nickel–tungsten sulfide catalyst (*13*). The writer is not sure that the process is used commercially. Even so, the reader may find useful the large amount of data and correlations presented.

A. Regeneration

The Pd catalysts can be regenerated with air and steam, using the usual "hot spot" precautions. Burning starts about 400°C and the maximum is 525°C.

B. Suppliers of Proven Catalysts

Catalysts and Chemicals, Inc.

C32–2: Pd–Cr on a specially prepared alumina support, as 4–8 mesh spheres.

Catalysts Division of Chemetron Corporation

Girdler G–68: 0.5% Pd plus 0.5% Cr on alumina, as 3/16 × 1/8 or 3/16 × 3/16 inch tablets. The ABD is 1.1.

The nickel catalysts developed for this process are available only to licensees of the process. If the writer had a problem in this area and wanted to use a nickel catalyst he would start with Harshaw Ni–0301 T in reduced form. If this was not satisfactory, he would then go the the sulfided form. If neither of these worked, he would then go the nickel tungsten, Harshaw Ni–4303 E, in the sulfided form.

Harshaw Chemical Company supplies:

Ni–0301 T: 11% Ni as oxide supported on activated alumina; as 1/8 inch tablets. The ABD is 1.1, the surface area 64 m^{2}gm^{-1}, and the pore volume 0.32 cm^{3}gm^{-1}.

Ni–4303 E: 6% Ni + 19% W as oxides on alumina; 1/12 extrudates. The ABD is 0.8, the surface area 152 m^{2}gm^{-1}, and the pore volume 0.54 cm^{3}gm^{-1}. H_2S at atmospheric pressure and 500°C sulfides the catalyst (*19*).

V. Hydrogenation of Fats and Oils

The hydrogenation of double bonds in fats and oils is one of the older commercial catalytic processes. It is used to convert oils or soft fats into lard or margarine. It is also used to convert them to hard fats, stearin (*11*, *28*). In "selective" hydrogenation, the objective is to convert the linoleates, linolenates, or similar polyunsaturated glycerides into oleates or more precisely into "isooleates" since the remaining double bond is not in the cis-9 position as in oleic acid. This involves hydrogenating one or two double bonds and leaving one. In nonselective hydrogenation, all double bonds may be saturated.

There is an inherent tendency for one of a multiple double bond fat acid ester to hydrogenate before the last one. Selective Ni catalysts are available to aid in this selectivity. With an active slurried catalyst in a stirred reactor, selectivity is favored by very low concentrations of hydrogen on the surface of the catalyst, i.e., by low H_2 pressure, low stirring rates, and high temperature (*24*). Conversely, the nonselective reaction is favored by low temperature, higher pressure, and good agitation. These differences show up clearly in the following examples of operating conditions, using the same active nickel catalyst.

Selective operation: Temperature 180°C, pressure 1.4 atm, and 0.1% Ni catalyst.

Nonselective operation: Temperature 125°C, pressure 3.4 atm, and 0.1% Ni catalyst.

Under highly selective conditions, the linoleate hydrogenation can be fifty times the rate of oleate hydrogenation, while under nonselective conditions it might be only fourfold.

Increasing temperature not only affects selectivity but also increases reaction rate, as would be expected. Similarly, pressure also increases the reaction rate. Within limits, increasing catalyst concentration increases the reaction rate. Mainly, increased catalyst concentrations are used to over come the deleterious materials in raw or only partly refined oils. For a refined oil the catalyst concentration is within the range of 0.01–0.02% Ni based on the oil.

The finished oil is filtered to remove the catalyst (with added filter aide if none was present in the original catalyst). The catalyst is not spent and is sometimes reused by discarding a part and adding some fresh. Or the used catalyst may be employed on a raw or hard-to-hydrogenate oil under more severe conditions. More often, the used catalyst is discarded and only fresh catalyst used, so that predictable results can be had from batch to batch.

Several types of stirred batch reactors are used. One is a "dead end" reactor, partly filled with oil, into which hydrogen is fed to maintain a predetermined pressure. A bottom stirrer keeps the catalyst in suspension while the top stirrer mixes hydrogen from the dead space into the oil. A second type is a vertical cylinder fitted with a "spider" at the bottom of the reactor for distributing the hydrogen. Agitation is provided by pumps that circulate oil, hydrogen, or both. External cooling may be used to control the temperature. Continuous processes using fixed catalyst beds are not common.

The batch size varies between 20 and 30 tons. It is possible to complete a batch in 1 to 2 hours with about half the time being required for actual hydrogenation.

A. Poisons

CO is a temporary poison, presumably because it combines with the Ni to form a surface carbonyl. The poisoning disappears when the source to CO is removed. CO may come from the method of preparing the H_2.

Besides sulfur and halogen compounds, phosphatides, gums, and soaps act as poisons in fat hydrogenations.

B. Suppliers of Proven Catalysts

Catalysts Division of Chemetron Corp.

G–15: an electrolytically precipitated nickel supported on kieselguhr (a filter aide) and suspended in a hardened edible oil or tallow. It is about 25% Ni, 10% kieselguhr, and 65% protective medium. It is supplied especially for use with highly refined oils.

G–53: about 25% Ni, 15% kieselguhr, and 60% protective medium, is supplied for the more difficultly hydrogenated oils.

G–70: about 25% Ni, 1% Zr, 14% kieselguhr, and 60% protective medium; as granules passing a 6 mesh screen. As supplied, it is active for low temperature use and is reusable for elevated temperature hydrogenation of the more difficult-to-hydrogenate oils (e.g., tallows and marine oils).

Harshaw Chemical Company

Nysel: a catalyst for selective hydrogenation containing 25–30% active nickel on a specially prepared support and protected by 60% stearin.

Rufert Catalyst: finely divided pure nickel dispersed in and protected by stearin (24–26% Ni). No filter aide is included so two parts of filter aide should be added for each part of Ni before filtering.

Nickel Formate: This can be used to prepare an active catalyst in the oil. Finely divided nickel formate dihydrate is suspended in the oil and heated with mixing. When decomposition starts at about 150°C, hydrogen is slowly bled through the oil. Over a period of about 2.5 hours the temperature is brought to 240°C where it is held for 1.5 hours (*30*). The catalyst is ready for use. This technique is used to make a master batch of catalyst from which portions are used for the individual hydrogenations.

VI. Hydrogenation of Aromatic Hydrocarbons

The hydrogenation of benzene to cyclohexane and naphthalene to tetralin or decalin are the main commercial aromatic hydrocarbon hydrogenations. Almost any of the Ni and Pt catalysts active for hydrogenating olefins to paraffins can be used.

A. Benzene

Benzene can be hydrogenated by nickel catalysts in a fixed bed in the 65°–250°C temperature range, 5–35 atm pressure, and LHSV 1–4 $V_l V_c^{-1}$ hr^{-1}. The reaction is strongly exothermic, so the temperature must be controlled by removing the heat. Otherwise, the high temperature favors incomplete conversion of the benzene. Many specifications call for 99.9% cyclohexane, so special care must be taken to insure essentially complete hydrogenation.

One way of controlling the temperature is to use a relatively inactive catalyst so that the heat of reaction is spread through a large part of the bed. Another method is to cool and recycle some of the cyclohexane produced. Alternatively, the catalyst can be put in the tubes of a heat exchanger with boiling water outside the tubes.

Still another method carries out the hydrogenation in the liquid phase in a stirred reactor combined with a fixed bed (vapor phase) reactor to complete the hydrogenation (*3*, *7*). Raney nickel is used at 200°C and 15 atm pressure. Hydrogen and benzene (liquid) are continuously fed to the reactor. Vapor (98–99+% cyclohexane) is continuously removed and, with hydrogen, is fed to a fixed bed of Ni catalyst to hydrogenate the last benzene. The heat of vaporization of the cyclohexane in the stirred reactor is used to remove the heat of reaction. Note that the catalyst in the stirred reactor acts as a "guard case," removing poisons from the vapor so that a poison-free feed can go to the second stage, where catalyst activity must be kept high. The second reactor is not cooled. When conversion in the first reactor drops to 95% cyclohexane, the second reactor starts to heat, indicating a need for new catalyst in the stirred reactor. The final product contains about 20 ppm of benzene.

Stirring in the first reactor is obtained by removing liquid plus catalyst from the bottom of the reactor and pumping it to the top. This circulating stream is cooled to help remove heat of reaction.

B. Naphthalene

Naphthalene hydrogenation occurs in two readily separable stages: first to tetralin and then to decalin. The same catalysts that hydrogenate benzene also hydrogenate naphthalene and tetralin under essentially comparable conditions. Actually, hydrogenating to tetralin is easier than hydrogenating benzene. The rate of naphthalene hydrogenation is roughly ten times as fast as the rate of tetralin to decalin. It is thus easy to hydrogenate most of the naphthalene to tetralin but difficult to do it without producing some decalin.

Benzothiophene is a common impurity in naphthalene and this will poison the more active nickel and platinum catalysts. The "cobalt molybdate on alumina" type catalysts can be used on such stocks at 225°–350°C and at least 30 atm pressure. Preferably, the catalyst is sulfided first and it may be necessary to add a little H_2S to the hydrogen stream to maintain the catalyst activity if there is not enough sulfur in the feed.

C. Suppliers of Proven Catalysts

Most of the Ni and Pt catalysts are listed in Section II. The following ones, as well as Raney nickel, are suggested for hydrogenating aromatic hydrocarbons.

Catalysts and Chemicals, Inc.
C46–5: nickel oxide on alumina in the form of tablets or extrusions.

Catalysts Division of Chemetron Corp.
G–33 and G–52.

Harshaw Chemical Company
Ni–0101 T, Ni–0102 T, Ni–0104 T, Ni–0109 T, Ni–0116 T, and Ni–1404 T.
Ni–1406 T, 46% Ni (unreduced) on a special support; 5/32 and 7/32 inch tablets. The ABD is 1.3.

Universal Oil Products Co.
H–2, H–3, and H–4.

VII. Hydrogenation of Nitroaromatic Hydrocarbons

The hydrogenation of nitrobenzene to aniline is the major commercial nitroaromatic hydrogenation, with the hydrogenation of *m*-dinitrotoluene to *m*-toluylene diamine probably being second. There is one important safety precaution to be observed when considering the hydrogenation of a nitro-compound. The reaction is so strongly exothermic that a runaway hydrogenation can (and did) result in an explosion. Adding a nitro compound and catalyst to an autoclave (or bomb) and charging with high pressure hydrogen is a hazardous procedure and must be avoided. The supply of nitro compound or hydrogen to the reaction should be kept under control at all times.

Almost any hydrogenation catalyst worthy of the name will hydrogenate nitrobenzene. Copper is most commonly used commercially for two reasons: (*a*) Its low activity spreads the reaction zone in a fixed bed and so helps control the exothermic reaction; (*b*) copper catalysts do not readily hydrogenate the aromatic ring so that aniline is the product instead of cyclohexylamine.

One procedure is described as follows (*10*). One kilogram of nitrobenzene per cubic meter of circulating hydrogen is carried to a fixed bed of copper catalyst at 200°C. The heat of reaction is sufficient to raise the exiting products to 350°C. The aniline is recovered and the unused (cold) hydrogen is recycled to the reactor.

Alternatively, the reaction can be conducted with powdered catalyst suspended in the liquid in a stirred reactor very much as the fats and oils are hydrogenated (Section V). This technique is applicable to *m*-dinitrotoluene, especially in a solvent. In this operation the rate of hydrogen addition, and hence the rate of hydrogenation and heat release, can be controlled to keep the reaction safe. Pd will work at atmospheric temperature and pressure. Raney copper and almost any of the reduced nickel or Raney nickel catalysts listed in Section II can be used. Temperature is in the 50°–150°C range, pressure 1–5 atm, and catalyst concentration around 0.1 to 1.0%.

A. Suppliers of Proven Catalysts

Davison Chemical Division of W. R. Grace & Co.

Grade 29: an active copper catalyst powder prepared from Raney copper (CuAl alloy).

Grade 2913: the Raney CuAl alloy that has to be activated with NaOH like Raney nickel (cf. Section II, E for procedure).

Harshaw Chemical Company

Cu–2501 G: 6% Cu as carbonate mounted on a silica support; 4–10 mesh granules. The ABD is 0.5, the surface area 4 m^2gm^{-1}, and the pore volume 0.61 cm^3gm^{-1}. This can be decomposed and reduced in place at 250°–300°C.

Chapter 15 lists suppliers of copper chromite catalysts which are also useful for hydrogenating nitro compounds in place of copper.

VIII. Hydrogenation of Aniline

Aniline is hydrogenated (*23*) to cyclohexylamine and, so far as the author is aware, always with some dicyclohexylamine as a byproduct. The addition of a little alkali, e.g., 0.1% of NaOH or CaO, to the reactor or to the catalyst partly inhibits dicyclohexylamine formation. Another technique recycles the dicyclohexylamine.

For the batch stirred reactor, Raney Ni, Raney Co, prereduced Ni on kieselguhr, or Ru on charcoal are all useful. Reaction conditions are: Temperature 150°–250°C, pressure 35–150 atm, with catalyst concentrations of 0.1–1.0%. Temperatures much above 250°C cause deamination to cyclohexane.

For fixed bed hydrogenation, Ru on alumina is a useful catalyst at 150°–250°C, 50–100 atm pressure, LHSV 0.5–2 $V_l V_c^{-1}$ hr^{-1} and H_2:aniline mole ratio 3–5.

A. Suppliers of Proven Catalysts

See Section II, E for suppliers of most of the catalysts mentioned.

Davison Chemical Division of W. R. Grace & Co.
- Grade 27: active Raney cobalt powder.
- Grade 2713: 50% Co–50% Al alloy powder, a Raney alloy that must be leached with NaOH before use. See Raney catalysts in Section II, E for leaching procedure and precautions.

IX. Hydrogenation of Pyridine and Quinolines

Ruthenium dioxide has been found to be a satisfactory catalyst for hydrogenating pyridine to piperidine, as well as for hydrogenating many of its homologs (*12*). Conditions for a batch stirred reactor are 90°–100°C and 70 atm pressure, with 1–2% RuO_2.

The same catalyst and conditions may be used for hydrogenating quinoline and isoquinoline, especially, to the *N*-tetrahydroquinolines which form first.

It should be noted that alcohols are not satisfactory solvents if these hydrogenations are to be carried out in solution, for *N*-alkylation occurs with the alcohol furnishing the alkyl group.

A. Suppliers of Proven Catalysts

Both Engelhard Industries and Johnson Matthey, Inc. supply RuO_2.

X. Hydrogenation of Phenol

Phenol may be hydrogenated to cyclohexanol with a number of catalysts. Some cyclohexanone is usually formed, especially when hydrogenation is not complete. Raney nickel or the prereduced nickel catalysts are satisfactory. Adding 0.1% of sodium phenoxide to a batch stirred reactor seems partially to inhibit cyclohexanone formation.

For a batch stirred reactor and prereduced nickel catalysts, useful conditions are 150°–250°C and 10–150 atm pressure, with 0.1–1% catalyst. For a fixed bed reactor, the following conditions work with a supported prereduced nickel catalyst: Temperature 150°–250°C, pressure 75–200 atm, LHSV 0.75–2, and H_2:phenol mole ratio 3–6.

A. Suppliers of Proven Catalysts

See Section II, C, D, and E for suppliers of nickel catalysts.

XI. Hydrogenation of Nitriles

Most often a primary amine is the desired product from the hydrogenation of a nitrile. There is a tendency however to form secondary and tertiary amines along with the primary amine, as well as other products. One of the more practical methods of minimizing the undesired products is to conduct the hydrogenation in the presence of ammonia. Both cobalt and nickel catalysts are used. The following conditions are suggested for a stirred batch reactor: 0.5% of powdered, activated (reduced) catalyst, H_2 partial pressure 18 atm, NH_3 partial pressure 8 atm, about 0.1% NaOH in the liquid, and 150°C. (Safety precautions: No copper or copper alloys are permissible in the presence of NH_3. Be sure that the Bourdon tube in the pressure gauge is a steel one.)

A. Adiponitrile to Hexamethylene Diamine

The technique described above, or a similar one, has been used to hydrogenate adiponitrile to hexamethylene diamine as one of the steps in a nylon manufacture.

B. Nitriles to Secondary Amines

If one wants the secondary amine instead of the primary one, the ammonia and NaOH should be omitted and the following conditions used:

Temperature 200°C, 3–7 atm hydrogen pressure, and 0.5% of powdered nickel catalyst.

Unsaturated nitriles (e.g., oleyl nitrile) can be converted to the unsaturated secondary amine using a copper chromite catalyst: Temperature 200°C, 15 atm hydrogen pressure, and 2–4% copper chromite powder.

C. Suppliers of Proven Catalysts

Catalysts Division of Chemetron Corp.

G–67: a zirconium promoted cobalt oxide catalyst specifically for nitriles to amines, as powder or tablets. It must be reduced before use.

G–67RS: the reduced and stabilized form of G–67. It is ready to use.

Davison Chemical Division of W. R. Grace & Co.

Grade 27: active Raney cobalt shipped in water for protection.

Grade 2713: a 50% Co–50% Al Raney alloy that must be NaOH leached for activation (see Section II, E for activation procedure).

Grade 28: active Raney nickel (shipped in water for protection).

Grade 2813: a 50% Ni–50% Al Raney alloy that must be leached with NaOH for activation.

Harshaw Chemical Company

Co–0101 T: 35% Co as oxide supported on kieselguhr, 3/16 inch tablets. Must be reduced before use. The ABD is 1.1, the surface area 130 m^2gm^{-1}, and the pore volume 0.35 cm^3gm^{-1}.

Co–0108 T: 39% Co on kieselguhr, activated and stabilized (ready to use); as 3/16 inch tablets. The ABD is 1.0, the surface area 140 m^2gm^{-1}, and the pore volume 0.40 cm^3gm^{-1}. The tablets should be powdered before use.

Ni–0104 P: a 58% Ni on kieselguhr, activated and stabilized in which 60–65% of the nickel is reduced (ready for use).

Rufert Nickel: 24 to 26% finely divided active nickel dispersed in stearin for protection. Although supplied primarily for hydrogenating fats and oils, this catalyst is suggested for converting nitriles to secondary amines.

For suppliers of copper chromite catalysts, see Chapter 15.

XII. Hydrogen Sulfide Synthesis

H_2S may be synthesized from the elements at about 500°C in yields of 95% or better. Bauxite, kaolin, and similar clays, or cobalt on alumina, are catalysts.

A. Suppliers of Proven Catalysts

Harshaw Chemical Company

Co–0301 E: 10% CoO on alumina; as 3/16 inch extrudates. The ABD is 1.2, the surface area 6.8 m^2gm^{-1}, and the pore volume 0.07 cm^3gm^{-1}.

Engelhard Minerals and Chemicals Corp.

Porocel: a 6–8 mesh bauxite.

REFERENCES

1. Adkins, H., "Hydrogenation." Univ. of Wisconsin Press, Madison, Wisconsin, 1937.
2. Anonymous, *Petrol. Refiner* **39,** No. 9, 252 (1960).
3. Anonymous, *Hydrocarbon Process.* **46,** No. 11, 164 (1967).
4. Bond, G. C., "Catalysis by Metals." Academic Press, New York, 1962.
5. Cohn, J. G., *Engelhard Ind. Tech. Bull.* **9,** No. 1, 5 (1968).
6. De Malde, M., Di Cio, A., and Mauri, M. M., *Hydrocarbon Process.* **43,** No. 7, 149 (1964).
7. Dufau, F. A., Eschard, F., Haddad, A. C., and Thonon, C. H., *Chem. Eng. Progr.* **60,** No. 9, 43 (1964).
8. Ellis, C., "Hydrogenation of Organic Substances." Van Nostrand, New York, 1930.
9. Emmett, P. H., ed., *Catalysis* **3** (1955); **4** (1956); **5** (1957).
10. Faith, W. L., Keyes, D. B., and Clark, R. L., "Industrial Chemicals," p. 118. Wiley, New York, 1957.
11. Feuge, R. O., *Catalysis* **3,** 413 (1955).
12. Friedlander, M., and Stone, G. R., *J. Org. Chem.* **26,** 3805 (1961).
13. Hoffman, E. J., Lewis, E. W., and Wadley, E. F., *Petrol. Refiner* **36,** No. 6, 179 (1957).
14. Kelly, A. W., Hurwitz, H., and Gamero, A., *Am. Chem. Soc. Div. Petrol. Chem. Preprints* **11,** No. 1, 123 (1966).
15. Kirk, M. C., and Reif, H. E., *Chem. Eng. Progr.* **61,** No. 3, 64 (1965).
16. Lester, R., *Hydrocarbon Process.* **40,** No. 9, 175 (1961).
17. Lieber, E., and Morritz, F. L., *Advan. Catalysis* **5,** 417 (1953).
18. Lindlar, H., *Helv. Chim. Acta* **35,** 446 (1952).
19. Massoth, F. E., and Bidlack, D. L., *Am. Chem. Soc. Div. Petrol. Chem. Preprints* **14,** No. 1, 107 (1969).
20. Peterson, H. C., and Jamieson, G. P., *Am. Chem. Soc. Div. Petrol. Chem. Preprints* **7,** No. 3, 23 (1962).
21. Rapoport, I. B., "The Chemistry and Technology of Synthetic Liquid Fuels," TP 343 R 18. Natl. Sci. Found., Washington, D.C.
22. Rylander, P. N., "Catalytic Hydrogenation over Platinum Metals." Academic Press, New York, 1967.
23. Rylander, P. N., see reference 22, p. 352.
24. Satterfield, C. N., and Sherwood, T. K., "The Role of Diffusion in Catalysis," p. 52. Addison-Wesley, Reading, Massachusetts, 1963.

25. Schmauch, G. E., and Singleton, A. H., *Ind. Eng. Chem.* **56,** No. 5, 20 (1964).
26. Smith, H. A., *Catalysis* **5,** 210 (1957).
27. Storch, H. H., "The Fischer-Tropsch and Related Syntheses." Wiley, New York, 1951.
28. Waterman, H. I., "Hydrogenation of Fatty Oils." Elsevier, New York, 1951.
29. White, P. T., Porter, F. W. B., and Yeo, A. A., *Proc. 5th World Petrol. Congr. New York, 1959* **3,** 335 (1960).
30. Wurster, O. H., *Ind. Eng. Chem.* **32,** 1193 (1940).

CHAPTER 14

SYNTHESIS OF AMMONIA AND METHANOL

These two commercial operations will be treated together for two reasons: (1) Both, as of this writing have been undergoing immense changes in technology resulting from the improvements in synthesis gas production (see Chapter 11) and turbo-compressors (2). The plants and their operation are similar in many respects.

I. New Technology

Until recently, most plants of both processes operated at about 300 atm pressure. As a result, the plant was an enormous compressor house filled with huge reciprocating compressors. The catalytic reactor was a minor piece, in size, of the equipment. The capacity of a plant was about 300 tons per day.

In contrast, in the current plants, the batteries of reciprocating compressors have disappeared and been replaced by a centrifugal compressor. Two things have brought about this change: (1) The technology to operate the synthesis gas plant at pressures approaching 30–35 atm instead of 5–10 atm (see Chapter 11). (2) The development of centrifugal compressors capable of compressing this gas to 150–200 atm (*3*). In both ammonia and methanol syntheses, the conversion per pass has decreased as a result of the lower pressure. However, with centrifugal recycle pumps, the cost of recycle decreases, so the drop in conversion per pass is acceptable.

II. Ammonia Synthesis

Ammonia as NH_3 must be made from $3\ H_2 + N_2$. However, kinetic studies with plant data indicate that the ammonia production rate can be increased if the synthesis gas contains somewhat more N_2, e.g., 2.5 H_2:N_2 (*6*). The mixture of nitrogen and hydrogen is passed over a specially prepared iron catalyst in the temperature range of 450°–520°C and a pressure range of 140–1000 atm, at a GHSV of 10,000–55,000 $V_g V_c^{-1}$ hr^{-1}. The reaction is exothermic and the reactor is constructed with heat exchangers inside it, so the heat of reaction can be used to preheat the incoming gases. Actually, the cool incoming gases pass along the inside wall of the pressure vessel to keep it cool before going to the catalyst held in vessels inside the pressure vessel. The reaction si an equilibrium reaction and the reacents only approach equilibrium before they are removed from the reactor. Conversion per pass is 50–80% of equilibrium. The exit gases are cooled, the liquid ammonia separated and the unconverted gases recycled. The gas always contains some methane and argon, and possibly some other inerts, so a part of it is vented before the rest of the gas is recycled. The ammonia production rate is about 30 kilograms per liter of catalyst per day.

In a plant operating at about 170 atm, there is around 13.5% NH_3 in the reactor effluent (*5*), This is removed as a liquid first by water cooling and then by refrigeration. In plants operating at pressures higher than 170 atm the concentration of ammonia in the reactor effluent may be higher so water cooling may be sufficient to remove an adequate amount of ammonia before recycling.

With the pressures used in various plants ranging all the way from 140 to 1000 atm, one must presume that at the time a given plant was built, the pressure used produced ammonia at the lowest cost for that size plant. As this is written, it is indicated that about 140 atm is the most economical pressure for plants with capacities of 600–1000 tons day^{-1}. For a 1500 ton day^{-1} plant, 200 atm is considered more economical (*8*).

A. Catalysts

The synthetic ammonia catalyst has probably had more work done on it than any other catalyst for any other chemical (*2, 4*). As a result, one may become hopelessly confused in trying to follow the literature on it. In general, the catalyst is a magnetite type of iron oxide that has been promoted with small quantities of potassium compounds, alumina with or without magnesia, and sometimes silica. The following illustrates one composition.

Fe_2O_3	60.7%
FeO	34.0%
K_2O, Al_2O_3, CaO	4.9%
SiO_2	0.3%

Actually, this is merely an indication of the raw material that is put into the reactor for forming the catalyst by reduction.

B. Catalyst Reduction

Catalyst suppliers provide a reduction procedure for their catalysts. The following illustrates one reduction procedure. The reactor is equipped with electrical heaters to provide heat to get the reaction started.

The catalyst is reduced by ammonia synthesis gas circulated at about 140 atm pressure. Heat the catalyst bed at a rate of 25°C hr^{-1} until 300°–350°C is reached when reduction starts. There will be a "hot spot" in the catalyst as the catalyst is reduced and this can be followed with a thermocouple to determine reduction progress. As the catalyst is reduced, ammonia formation will begin and aqua ammonia of increasing ammonia concentration will collect in the separator. By the time the "hot spot" reaches 450°C, the heat of ammonia formation dominates and the rate of temperature rise which has been slow will tend to increase rapidly as the last catalyst is reduced. The temperature is kept under careful control as it and pressure are brought to operating conditions.

The catalyst must be sturdy so that the reduction and use do not cause attrition, thus forming fine particles and increasing the pressure drop across the bed.

C. Poisons

One of the purposes of the promotors in the catalyst is to attenuate the effect of poisons. CO, CO_2, O_2, and H_2O are all catalyst poisons and are usually present to some degree in the synthesis gas as a result of the way it was prepared. Chlorides and sulfur compounds are also poisons, but these are rarely present in synthesis gas prepared by current methods.

The life of the catalyst depends on the severity of the operation and exposure to poisons. A life of 3–5 years is expected in most plants although it may be as short as 3–4 months in 1000 atm plants. The spent catalyst is discarded as it is not practical to regenerate it.

D. Suppliers of Proven Catalysts

Catalysts and Chemicals, Inc.

C73–1 and C73–2. Both are promoted iron oxides and are furnished as 6 × 12 mm granules. C73–2 contains a higher concentration of promotors and is suggested for the initial 10% of the catalyst bed, with C73–1 making up the rest.

Catalysts Division, Chemetron Corporation supplies a promoted iron oxide, LN–II.

Halder Topsoe, Inc. supplies promoted iron oxide catalysts, KM–I and KM–II.

Katalco Corporation

ICI 35–4: a promoted iron oxide. The ABD is 2.6, the surface area (after reduction) 10–15 m^2gm^{-1}, and the pore volume 0–0.1 cm^3gm^{-1}.

There are other suppliers who make their catalysts available to process licensees.

III. Methanol Synthesis

Synthetic methanol is made by passing a mixture of CO and/or CO_2 + hydrogen over a "zinc chromite" catalyst. Operating conditions are as follows. Temperature 300°–400°C, pressure 100–600 atm, H_2:CO ratio 2.1–2.5:1, GHSV 20,000–60,000 $V_gV_c^{-1}$ hr^{-1}, with a conversion per pass of about 5%. Conversion is kept below equilibrium to minimize byproduct formation.

The reaction is strongly exothermic so that provision must be made to carry this heat away. As a result, internal heat exchangers of the type common for synthetic ammonia are used. In fact, a given reactor may be used for either ammonia synthesis or methanol synthesis by changing catalyst and feed gases. However, with methanol, there must not be any carbon steel in the reactor; otherwise it may disappear as the result of volatile iron carbonyl formation. Stainless alloys are used (*7*).

As has been mentioned, plants have been designed operating at lower pressures than those indicated above, taking advantage of the economies offered by centrifugal compressors. Even further evolutionary changes in the methanol process involve a more active catalyst based on copper. Copper catalysts, and especially zinc–copper chromites, have been known for many years and their superior activity well established (*7*). However, the activity was lost rapidly so the more rugged zinc chromites have been used.

The more active catalysts have permitted plant design at pressures as low as 50 atm and temperatures of 230–270°C (*9*). In Chapter 11, it was pointed out that the steam reforming of natural gas produced a gas with an H_2O:CO ratio of about 3 when one of about 2 is needed for methanol synthesis. It was also indicated that an external source of carbon, e.g., as CO_2 was often used to complete the stoichiometric needs of the process. Using the new catalysts, there are circumstances where it is apparently economical to use the 3:1 H_2:CO ratio gas in the reactor and take advantage of the mass action effect of the extra hydrogen on the equilibrium to favor methanol production. The excess hydrogen is vented and used as fuel.

A. Byproducts

Although the reaction seems simple enough, small amounts of byproducts are formed that must be removed. The low boiling ones are dimethyl ether and formaldehyde, while the high boiling ones are higher alcohols, i.e., isopropanol, isobutanol and still higher boiling alcohols.

Dimethyl ether and the higher boiling alcohols are removed by fractionation. Formaldehyde is removed by oxidizing it with permanganate.

If an even larger proportion of higher alcohols is wanted, adding alkali, especially K_2CO_3 or KOH, to the catalyst is effective. The catalyst is then used at lower space velocities and somewhat higher temperatures (*1*). Adding olefins to the feed gases also increases the yields of higher alcohols (*7*).

B. Poisons

Iron or nickel in the catalyst acts as an independent catalyst, converting CO into CH_4 and so wasting feed. When present, they are usually lost as a volatile carbonyl and removed from the system. H_2S, COS, or CS_2 act as temporary poisons with zinc catalysts converting zinc to zinc sulfide. Once the sulfur compounds are removed from the feed, the catalyst recovers via the reaction:

$$ZnS + H_2O = ZnO + H_2S$$

with H_2S gradually working itself out of the system. H_2S is to be avoided because, through the reaction:

$$CH_3OH + H_2S = CH_3SH + H_2O$$

it leads to a product with poor odor.

The catalysts containing copper are damaged by sulfur compounds with the copper being irreversibly converted to the sulfide.

C. Suppliers of Proven Catalysts

Catalysts and Chemicals, Inc.

C70–2: a zinc chromium oxide catalyst as ¼ × ⅛ inch tablets. It must be reduced before use.

Harshaw Chemical Company

Zn–0302 T: 74% ZnO + 23% chromium oxide with the chromium in the hexavalent state, as 3/16, ¼, or ⅜ inch tablets. Must be reduced before use.

Zn–0308 T: the reduced form of Zn–0302 T containing 22% chromium (trivalent) oxide, as ¼ inch tablets.

Zn–0311 T: similar to Zn–0302 T but may give a little different selectivity, as ³⁄₁₆, ¼, or ⅜ inch tablets.

Zn–0312 T: similar to Zn–0308 T but may give a little different selectivity, as ³⁄₁₆, ¼, or ⅜ inch tablets.

There are other methanol catalysts, including the copper base ones, which are available only to licensees of the process, e.g., Katalco Corporation supplies 51–1, a copper oxide on a support, on this basis.

During reduction, the "zinc chromite" catalysts shrink. If a full reactor of reduced catalyst is wanted, reduction is done in a separate vessel. Reduction (H_2 + CO) begins at about 175°C and is completed as the temperature is raised gradually to 300°C.

REFERENCES

1. Anderson, R. B., Feldman, J., and Storch, H. H., *Ind. Eng. Chem.* **44,** 2418 (1952)
2. Bokhoven, C., van Heerden, C., Westrik, R., and Zwietering, P., *Catalysis* **3,** 265 (1955).
3. Finneran, J. A., and Masur, P. J., *Oil Gas J.* **67,** No. 29, 70 (1969).
4. Frankenburg, W. G., *Catalysis* **3,** 171 (1955).
5. Gambrell, L. M., Bradley, L. D., and Aalund, L. R., *Oil Gas J.* **65,** No. 13, 72 (1967).
6. Hays, G. E., Poska, F. L., and Stafford, J. D., *Chem. Eng. Progr.* **60,** No. 1, 61 (1964).
7. Natta, G., *Catalysis* **3,** 349 (1955).
8. Quartulli, O. J., Fleming, J. B., and Finneran, J. A., *Hydrocarbon Process.* **47,** No. 11, 153 (1968).
9. Royal, M. J., and Nimmo, N. M., *Hydrocarbon Process.* **48,** No. 3, 147 (1969).

CHAPTER 15

HYDROGENATION OF CARBONYL GROUPS

For the simple hydrogenation of carbonyl groups many of the catalysts described in Chapter 13 will work. However, if other hydrogenatable groups are present in the molecule, it is useful to have catalysts that will hydrogenate the carbonyl group and leave the other group untouched. The "copper chromites" have such a selectivity in some molecules. This chapter summarizes the areas of utility and limitations.

In general, copper chromite hydrogenates the carbonyl group in aldehydes, ketones, and esters to the corresponding alcohols. The benzene nucleus resists hydrogenation, and the C=C double bond partly resists, while the carbonyl group is being hydrogenated. However, nitro, nitroso, cyano, oximino, or pyridinoid groups, if present, will be hydrogenated too (*1*).

I. Fats and Oils

The copper chromite catalysts are used to hydrogenate fats and oils to the corresponding alcohols. Typical operating conditions are 150°–350°C and 100–300 atm pressure. The operation is usually a batch process with the finely divided catalyst suspended in the reaction mixture. If the original oil contained carbon–carbon double bonds, these mostly survive but some are hydrogenated. Many "promotors" have been used with the copper chromite to hinder hydrogenation of the double bond without hurting activity for the carbonyl group. Cadmium oleate or carbonate seems to give some benefit as do cobalt compounds. However, the ones used commercially are proprietary.

II. Other Esters

The aliphatic esters behave like the fats and oils. The dibasic acids give glycols, e.g., the succinates give tetramethylene glycol, the adipates give hexanediol-1,6, etc.

The esters of aromatic acids tend to go all the way to the hydrocarbon when the carbonyl group is attached to the nucleus, e.g., ethyl benzoate can go to toluene. It is important to use an active, prereduced catalyst at a temperature in the 40°–110°C range to get the alcohol (*2, 8*). If the catalyst requires a temperature in the 150°–250°C, the hydrocarbon is the probable product.

The salicylate esters behave abnormally in that the benzene ring is hydrogenated; for example, methyl salicylate hydrogenates to 2-methylcyclohexanol. Presumably the salicylate reacts in the keto form which does not contain an aromatic nucleus (*1*).

III. Carbon–Carbon Double Bonds

Copper chromite will catalyze the hydrogenation of carbon–carbon double bonds but it is less active than nickel and is therefore not usually recommended when this is the sole activity desired. However, when the double bond is attached to an aromatic nucleus, nickel catalysts will tend to hydrogenate both the carbon–carbon double bond and the nucleus. Here copper chromite is useful, as only the carbon–carbon double bond is hydrogenated. For example, styrene gives ethylbenzene.

In physical mixtures containing both olefinic and aromatic hydrocarbons, the olefins can be selectively hydrogenated, leaving the aromatics.

It must be remembered that in the polycyclic aromatic hydrocarbons, not all rings are benzene rings. For example, phenanthrene gives 9,10-dihydrophenanthrene with copper chromite, leaving two benzene rings unhydrogenated (*1*).

Even though phenol is not hydrogenated with copper chromite catalysts, β-naphthol is, and it gives 1,2,3,4-tetrohydro-2-naphthol (*5*). Similarly, pyridine is not hydrogenated but quinoline gives 1,2,3,4-tetrohydroquinoline.

IV. Aldehydes and Ketones

Huge quantities of aldehydes are produced in the Oxo (Hydroformylation) process. These are hydrogenated by means of cobalt catalysts. Details and catalyst suppliers are given in Section I–H, of Chapter 21.

The active, prereduced copper chromite will reduce aldehydes and ketones at room temperature and 200 atm pressure, although the carbonyl group in aceto aromatics requires up to 100°C (*2*, *8*). With less active catalysts, higher temperatures are required for both types of carbonyl groups.

Acrolein gives some allyl alcohol but considerable *n*-propanol (*2*) even with the most active catalyst and the mildest conditions.

The major industrial aldehyde reduced using copper chromite is furfural. In furfural, the furan nucleus behaves like benzene and is not hydrogenated. Furfuryl alcohol is the main product (*4*). The reduction can proceed to methylfuran. If this is desired as the main product, it can be made at 200°–230°C and atmospheric pressure (*3*). Alternatively, at 100–200 atm pressure and 175°C, furfural gives 1,2- and 1,5-pentandiols (*1*).

V. Amides

Copper chromite catalysts can be used to hydrogenate the carbonyl group in amides to give amines. Like other reactions involving making amines, the yield of primary amine may be reduced by the reaction:

$$2\ RNH_2 = R_2NH + NH_3$$

Imides also are hydrogenated. For example, succinimide and glutarimide give pyrrolidine and piperidine, respectively. The *N*-alkyl amides and imides react similarly.

With aromatic amides, there is some tendency to go all the way to the hydrocarbon, e.g., benzamide to toluene. With active prereduced catalyst and reaction conditions such that partial rather than complete conversion is obtained, the amine is produced.

VI. Catalyst Activity

Most copper chromite catalysts are prepared by the thermal decomposition of an ammonium copper chromate. This may or may not be followed by an acetic acid leach. Catalyst which has not been heated above 350°C is amorphous to X-ray diffraction (*7*). Heating above 350°C forms crystalline copper oxide + copper chromite; the mixture is still a catalyst but is less active than the amorphous form. "Copper chromite" really describes or covers a variety of catalyst compositions as the list at the end of this chapter shows. Some of them contain barium and this stabilizes the catalyst so that activity is maintained longer.

Prereduced catalyst may be prepared in several ways. In the laboratory, it may be suspended in methanol, charged to a bomb, pressured with hydrogen to 250 atm, and heated to 100°C for 5 minutes. The bomb is then cooled, the hydrogen vented, and the material to be hydrogenated added to the methanol catalyst slurry. The hydrogenation is then carried out as usual (*2*).

For fixed bed use, the catalyst may be prereduced by heating in a stream of H_2 at 250 to 300°C and atmospheric pressure. The reaction is exothermic and in a fixed bed of catalyst a hot spot will pass through the catalyst. The temperature of this hot spot should not exceed 325°C. After the hot spot has passed through the bed, the catalyst is ready to use. It should not be exposed to air.

A. Regeneration

Ordinarily, the catalyst is discarded after use and not regenerated.

VII. Poisons

Sulfur compounds, as well as organic chlorine, bromine, and iodine compounds, are permanent poisons, since they convert the copper to the sulfide or halide. In many organic fluorine compounds, the fluorine is held tightly enough so that the compound can be hydrogenated successfully without losing fluorine to the catalyst (*6*). Carbon monoxide in the hydrogen acts as a temporary poison for hydrogenation in some reactions but the effect stops as soon as the carbon monoxide is removed.

VIII. Suppliers of Proven Catalysts

Calsicat Division, Mallinckrodt Chemical Works
: Copper chromite, CuO 39.6%, Cr_2O_3 41.7% and BaO 8.6% as powder. The ABD is 0.21 and the surface area 45–50 m^2gm^{-1}.

Catalysts Division of the Chemetron Corporation
: Girdler G–13 as a fine black powder.
: Girdler G–22: barium promoted as powder or 3/16 × 1/8 inch tablets.

Harshaw Chemical Company
: Cu–0202 P: 82% CuO + 17% Cr_2O_3 as powder. The ABD is 0.9, the surface area 10–13 m^2gm^{-1}, and the pore volume 0.55 cm^3gm^{-1}.
: Cu–0203 T: 78% CuO + 20% Cr_2O_3 as 1/8 inch tablets. A tabletted version of Cu–0202 P. The ABD is 2.2.

Cu–0401 P: 41% CuO, 44% Cr_2O_3, and 11% BaO as powder. The ABD is 0.9, the surface area 14–15 m^2gm^{-1}, and the pore volume 0.46 cm^3gm^{-1}.

Cu–0402 T: 35% CuO, 38% Cr_2O_3, and 10% BaO, as ⅛ inch tablets. The ABD is 1.6. A tabletted version of Cu–0401 P.

Cu–1106 P: 39% CuO, 43.5% Cr_2O_3, and 10% BaO, as powder. The ABD is 0.5, the surface area 50 m^2gm^{-1}, and the pore volume 0.76 cm^3gm^{-1}.

Cu–1107 T: 33% CuO, 38% Cr_2O_3, and 9% BaO, as ⅛ inch tablets. The ABD is 1.8. A tabletted version of Cu–1107 P.

Cu–1110 P: 41% CuO, 46% Cr_2O_3, and 9% BaO, as powder. The ABD is 0.65, the surface area 40 m^2gm^{-1}, and the pore volume 0.92 cm^3gm^{-1}.

Cu–1402 P: CuO 40%, Cr_2O_3 60%, as powder. The ABD is 1.1, the surface area 30 m^2gm^{-1}, and the pore volume 0.69 cm^3gm^{-1}.

Cu–1407 T: 37% CuO and 52% Cr_2O_3 as ⅛ inch tablets. The ABD is 1.7. A tabletted version of Cu–1402 P.

Cu–1800 P: 51% CuO and 47% Cr_2O_3, as powder. The ABD is 0.65, the surface area 40 m^2gm^{-1}, and the pore volume 0.76 cm^3gm^{-1}.

Cu–1808 T: 42% CuO and 38% Cr_2O_3, as ⅛ inch tablets. The ABD is 1.3. A tabletted version of Cu–1800 P.

REFERENCES

1. Adkins, H., "Reactions of Hydrogen." Univ. of Wisconsin Press, Madison, Wisconsin, 1937.
2. Adkins, H., Burgoyne, E. E., and Schneider, H. J., *J. Am. Chem. Soc.* **72,** 2626 (1950).
3. Burnette, L. W., Johns, I. B., Holdren, R. F., and Hixon, R. M., *Ind. Eng. Chem.* **40,** 502 (1948).
4. Calingaert, G., and Edgar, G., *Ind. Eng. Chem.* **26,** 878 (1934).
5. Dauben, H. J., Jr., McKusick, B. C., and Mueller, G. P., *J. Am. Chem. Soc.* **70,** 4179 (1948).
6. Renoll, M. W., *J. Am. Chem. Soc.* **68,** 1159 (1946).
7. Stroup, J. D., *J. Am. Chem. Soc.* **71,** 569 (1949).
8. Williams, J. L. R., *J. Org. Chem.* **19,** 1205 (1954).

CHAPTER 16

HYDROREFINING

Hydrorefining is the improvement in quality of a product (mostly hydrocarbons) by reacting it with hydrogen in the presence of a catalyst. The improvement may be in odor, color, tendency to deposit gum, storage stability, burning quality, etc. It does this by removing combined sulfur, nitrogen, oxygen, olefinic, or diolefinic hydrocarbons, or by saturating aromatic rings. The treatment is applied to gasoline, naphtha, jet fuel, kerosine, home heating oils, diesel fuels, lubricating oils, shale oils, coal tars, products from tar sands, etc. The process is especially important for removing S from products to be burned and so control SO_2 air pollution.

The catalysts are based on cobalt or nickel molybdenum aluminum or nickel tungsten aluminum compositions in which the original catalysts are prepared as oxides. Both nickel molybdenum and nickel tungsten catalysts must be sulfided for proper activity. It is not usually necessary to sulfide the cobalt molybdenum catalyst, but this is suggested in laboratory work and in commercial operation if the feed is low in sulfur.

The operation is conducted in a fixed bed of catalyst in an adiabatic reactor. Processing conditions are usually within the range: Temperature 300°–400°C, pressure 30–40 atm, LHSV 1–5 $V_l V_c^{-1}$ hr^{-1}, with up to 10 moles of H_2 circulated with each mole of hydrocarbon. Depending on the boiling range of the feed and the operating conditions, the feed in the reactor may be all vapor, mixed vapor–liquid, or essentially all liquid. When liquid is present it trickles down through the catalyst so at least in part it is a trickle bed operation. For mild treatment, no hydrogen need be circulated, i.e., it is replaced only as it is consumed or leaves the reactor dissolved in the product. For more severe treatment the hydrogen is circulated.

In this process, reasonable care must be taken to prevent oxidation (peroxide formation) of the feed. This means that liquid in storage tanks must be protected by floating roofs or blanketed with an inert gas such as CO_2, flue gas, or natural gas. Otherwise, oxidation products build upon the catalyst, causing an early regeneration (*17*).

After passing through the reactor, the product is cooled, passed to a liquid–gas separator where unconsumed H_2 is separated for recycle. The separated liquid is passed to another vessel where H_2S and NH_3 from the S and N compounds are separated from the purified hydrocarbons.

I. Gasoline or Naphtha Processing

For reforming naphthas, most plants operate with Pt–alumina catalysts. These can tolerate considerable amounts of sulfur compounds in the feed, but experience has shown that they operate with much greater efficiency and for longer periods if the content of sulfur and basic nitrogen compounds is low (<10 ppm S and <1 ppm N). If there is little or no nitrogen compounds in the feed, cobalt molybdenum catalysts can be used. When nitrogen compounds are present, nickel molybdenum or nickel cobalt molybdenum catalysts are preferred. With a severe nitrogen problem, two stage treaters can be used with cobalt molybdenum in the first stage and nickel molybdenum in the second. In this way 600 ppm N can be reduced to >1 ppm. On the average, the nickel molybdenum catalysts operate 10°–20°C lower in temperature than cobalt molybdenum.

With these catalysts sulfur and nitrogen compounds are both acted on but sulfur, usually present in higher concentration, is acted on more rapidly than the nitrogen compounds. In some way the catalyst breaks the C—S and C—N bond and saturates the ends of the broken bonds with H. The hydrogen consumption is usually less than 0.1 mole $mole^{-1}$ of feed.

Sometimes the feed to a reformer contains naphtha made by the thermal coking of high boiling petroleum residues. These naphthas contain olefins as well as sulfur and nitrogen compounds. The olefins are saturated while the nitrogen and sulfur are being removed. This reaction is strongly exothermic so the heat release is usually controlled by mixing the product from coking with naphtha direct from crude oil. This operation requires extra hydrogen for the olefins. Consumption is usually less than 0.5 moles $mole^{-1}$ of total feed.

Sometimes it is desirable to improve the antiknock susceptibility to lead alkyls of gasolines from catalytic cracking and this is done by saturating the olefins while desulfurizing. This is usually applied to that part of the gasoline boiling below 100°C.

II. Coal Tar Distillates

The benzene, toluene, and xylenes in certain coal tars contain sulfur compounds, mostly thiophenes, that are not desired. These may be removed with concentrated sulfuric acid but this removes hydrocarbons at the same time and reduces the yield. The sulfur compounds can be removed with these catalysts. However these catalysts will hydrogenate some of the aromatic hydrocarbons and also reduce the yield. To prevent this the upper end of the temperature range, the lower end of the pressure range or both are used. Hydrogen consumption is less than 0.1 mole $mole^{-1}$ of hydrocarbons.

III. Jet Fuel and Kerosine

Aromatic hydrocarbons in these products cause smoking when they are burned, and this is undesirable in both fuels. It is necessary to hydrogenate these aromatics to cycloparaffins as well as to reduce their S and N content. The cobalt molybdenum and nickel molybdenum catalysts will hydrogenate aromatic hydrocarbons. To reduce the aromatic concentration to the low level desired, pressures in the 100–200 atm are needed.

An alternate method uses one or two stage desulfurization and denitrogenation, after which H_2S and NH_3 are removed. The aromatics are then hydrogenated by more active, but S sensitive, catalysts at 50–100 atm (*9*, *22*). Hydrogen consumption depends on aromatic content. Pd or Pt catalysts on charcoal would be suitable for this hydrogenation except regeneration by burning is desirable so Pd or Pt on active alumina or other noncombustible supports are used (see Chapter 13 for Pd and Pt catalysts).

IV. Home Heating Oil and Diesel Fuel

These two oils are processed in the same way and are treated primarily to reduce sulfur and thus reduce air pollution. Because of their boiling range, they are at least partly liquid in the reactor, so the trickle bed operation is used. The sulfur levels in the product are 0.1–0.3%, so treating conditions are not nearly so severe as in naphtha treating for reforming or jet fuel hydrogenation.

TABLE VIII

SULFUR REDUCTIONS BY HYDROREFINING HOME HEATING OILS

Sulfur (%)	
Original	Product
1.0	0.1
0.7	0.05
0.6	0.06
1.8	0.08

Table VIII illustrates sulfur reductions obtained with four home heating oils.

That nitrogen is more difficult to remove than sulfur with current catalysts is illustrated by the data on three different oils in Table IX. Hydrogen consumption is around 0.1 mole mole^{-1} of oil.

TABLE IX

HYDROREFINING TO REMOVE SULFUR AND NITROGEN SIMULTANEOUSLY

Sulfur (%)		Nitrogen (%)	
Original	Product	Original	Product
0.8	0.05	0.015	0.007
1.19	0.07	0.056	0.041
1.58	0.14	0.012	0.007

Occasionally the aromatic content of diesel fuel is reduced to improve the burning quality. The reduction sought is much less than with jet fuel, so the cobalt molybdenum and nickel molybdenum catalysts are used.

V. Hydrogenating Feed for Catalytic Cracking

If the feed for catalytic cracking contains considerable sulfur and especially nitrogen, removing these by hydrorefining can increase the gasoline yield in catalytic cracking, reduce coke formation, and result in gasoline containing less sulfur. This is particularly true for feed stock that has been made by coking petroleum residue (*2, 4, 21*).

These benefits come from two sources. Removal of the basic nitrogen compounds prevents them from depositing on the acidic cracking catalyst and inhibiting its action, as well as from forming coke on the catalyst. The second benefit comes from sulfur removal. While sulfur compounds do not significantly inhibit the acid action of the cracking catalyst in the way nitrogen compounds do, the sulfur compounds are more readily adsorbed by the catalyst than hydrocarbons.

In catalytic cracking, aromatic rings are not cracked. They pass through the catalyst or adhere to it and form coke. Thus, the unconverted oil leaving a catalytic cracker contains most of the polycyclic aromatic hydrocarbons originally present in the feed plus those that may have been formed. If the polycyclic aromatics can be partially saturated so they contain only one aromatic ring with the other rings hydrogenated, they are quite susceptible to catalytic cracking. The cobalt molybdenum, nickel molybdenum, and nickel tungsten catalysts are all capable of hydrogenating these polycyclic aromatics. Under the conditions given above, from 40 to 70% or more of the polycyclic aromatic hydrocarbons are hydrogenated. Even higher percentages can be hydrogenated by raising the pressure or severely limiting the space velocity. Ordinarily, the incremental benefits are not worth the extra cost (*2, 4, 21*).

As might be expected, hydrogenating aromatics consumes more hydrogen than just removing S and N, i.e., 2–3 moles $mole^{-1}$ of oil versus 0.1 or less.

VI. Tar Sands Products

In the process operating commercially as this is written, the bitumen separated from the tar sands is first coked (thermally) yielding a "coke distillate." This is fractionated into a naphtha, kerosine, and gas oil. Each of these is hydrorefined with the type of catalysts discussed here and for the purposes described. The naphtha unit operates at 50 atm and removes S and N compounds and saturates olefins. Hydrogen consumption is about 1 mole $mole^{-1}$ of naphtha. The kerosine and gas oil units operate at 100 atm and saturate part of the aromatic hydrocarbons in addition to removing S and N. Hydrogen consumption is 3–4 moles $mole^{-1}$ of feed (*3*).

VII. Lubricating Oils

The objectives of treating lubricating oils are to improve color, resistance of color to change with storage (color stability) and resistance to oxidation.

For many years, these have been obtained by treating the oil with an adsorptive clay and, in some cases, with concentrated sulfuric acid. A portion of the oil was left imbibed in the spent clay along with the impurities adsorbed. Hydrorefining more selectively removes the sulfur and nitrogen compounds responsible for poor color and poor stability, so the oil previously imbibed in the clay is retained (*11*, *12*, *15*, *18*). As of 1968, 70% of the lubricating oils were hydrorefined (*18*). Hydrogen consumption is about 0.1 mole $mole^{-1}$ of oil.

As the molecular weight of petroleum fractions increase, the sulfur and nitrogen compounds are more difficult to hydrogenate, liberating the S and N, using currently available catalysts. Further, the N most difficult to remove is the major source of the undesired characteristics in the lubricating oils. Hence, the hydrorefining plants for the lubricating oils are designed to operate in the more severe part of the operating range. Temperatures tend to be in the 400°C range and pressures in the 100 atm range. Some hydrocracking may occur and possibly some dehydrogenation of cyclohexane rings to aromatic rings, in addition to the usual N and S removal.

Nickel molybdenum is the currently preferred catalyst for removing nitrogen compounds. In 1968, 60% of the catalyst in this service was nickel molybdenum (*18*).

VIII. Coal Tar Oils

Low temperature carbonization of coal produces distillates (low temperature coal tar) which have been studied as petroleum supplements. These distillates contain far more oxygen than petroleum distillates and, being produced by thermal means, they contain considerable olefins and aromatics. To convert them into petroleum supplements, the N and S content must be reduced and, because of the phenolic nature of most of the oxygen compounds (tar acids), the O must be removed or its compounds made inocuous.

An interesting study, using powdered WS_2 as the catalyst in a stirred reactor with a distillate from Utah coal, showed that 80–90% of the S, N, and O could be removed at 400°C and 100 atm in 2 hours, using 10% by weight of catalyst (*19*).

A different Utah coal distillate was studied using a nickel molybdenum catalyst in a fixed bed. It was necessary to use higher pressures, 135–200 atm, to get reasonable reaction rates (LHSV about 1 $V_l V_c^{-1}$ hr^{-1}). Hydrogen circulation was about 50 moles $mole^{-1}$ of coal distillate. These are relatively severe conditions. Virtually all the S was removed along with some 80% of the N and 90% of the O (*23*). Hydrogen consumption was 3–4 moles $mole^{-1}$ of coal distillate.

IX. Shale Oil

Shale oil has been both studied and used as a supplement to petroleum. The problems of cleaning it up are similar to those described above for low temperature coal distillates. Recent studies (*5, 16*) have favored coking the shale oil first and then hydrorefining the coke distillate to remove sulfur, nitrogen, and oxygen compounds and saturate the olefins.

Acceptable petroleum supplements were made from the coke distillate in a two stage hydrorefining treatment. In the first stage, conditions were mild to saturate the olefins, give a reduction in O compounds, and allow some S removal. Conditions in the first stage were: 245°C (average), 35 atm, LHSV 0.89 $V_l V_c^{-1}$ hr^{-1}, with H_2 circulation of about 2.3 moles per mole of distillate. The catalyst was presulfided commercial cobalt oxide–molybdenum oxide–alumina, and hydrogen consumption was 0.8 moles per mole of feed. Conditions in the second stage were 435°C, 100 atm, LHSV 1.0 $V_l V_c^{-1}$ hr^{-1} with H_2 circulation of about 11 moles per mole of feed. The catalyst was presulfided commercial nickel tungsten alumina, and hydrogen consumption was about 3.3 moles per mole of feed (*16*). Table X summarizes the removal

TABLE X

HYDROREFINING A SHALE OIL–COKE DISTILLATE

	Original coke distillate (%)	After first stage (%)	After second stage (ppm)
S	0.54	0.49	60
N	2.0	2.0	917
O	1.2	0.83	—

of S, N, and O. With a total hydrogen consumption of some 4.1 moles per mole of coke distillate, some aromatic rings were saturated as well as the olefins.

A series of catalysts was studied for single stage hydrorefining the coke distillate from a shale oil. At pressures up to 70 atm an HF treated cobalt oxide–molybdenum oxide–alumina was found the better of the twelve studied (*5*). Neither presulfided nickel oxide–molybdenum oxide–alumina nor presulfided nickel oxide–tungsten oxide–alumina was among the twelve catalysts reported.

X. Removing Metal Compounds

It is known that arsenic and lead are potent poisons for the noble metals used in reforming naphthas. Occasionally, these do get into the feed to reformers.

Similarly, iron, nickel, and vanadium compounds are deleterious to the cracking catalysts in that they promote undesired formation of hydrogen and coke. Nickel and vanadium compounds are present in the high boiling fractions of many crude oils and sometimes get into the feed to the catalytic cracker.

TABLE XI

REMOVAL OF ARSENIC AND LEAD BY A HYDROREFINING CATALYST

Arsenic (ppb)		Lead (ppb)	
Original	Product	Original	Product
84	1	138	5
32	1	6	5

Interestingly, the cobalt oxide–molybdenum oxide, nickel oxide–molybdenum, or nickel oxide–tungsten oxide, all on alumina, have the ability to remove these metals, possibly by adsorption on the alumina. Table XI illustrates arsenic and lead removal by a cobalt oxide–molybdenum oxide–alumina catalyst. Table XII shows how most of the arsenic was concentrated towards the entrance end of the bed.

TABLE XII

DISTRIBUTION OF ARSENIC IN A USED HYDROREFINING CATALYST

Level	Arsenic (ppm)
Top of Bed	1650
5%	800
10%	250
15%	110
20%	100
25%	5

Processing these heavy black oils is difficult for known catalysts. The residues contain components that readily deposit on the catalyst as coke or form materials that deposit as coke. Furthermore, nickel and vanadium compounds, if present, are concentrated in the residue and deposit on the catalyst. The vanadium is especially undesirable because, on regeneration by burning, it is converted into a low melting oxide that fluxes the catalyst and deactivates it. In a fixed bed, most of the vanadium is found in the inlet part of the catalyst bed, while nickel is distributed pretty well throughout the bed. The greater stability of the nickel compounds compared to vanadium ones during hydrodesulfurization can be seen in the data given in Table XIII, summarizing studies made at different space velocities (*13*).

TABLE XIII

REMOVING METAL COMPOUNDS DURING HYDRODESULFURIZATION OF A PETROLEUM RESIDUUM

		Product		
	Original	LHSV 4	LHSV 2	LHSV 1
V (ppm)	39.3	17.8	3.8	0.1
Ni (ppm)	2.5	1.6	1.0	0.6
Metals removed (%)	0.0	54	88	98

If the metal impurities are adsorbed in the pores of the alumina base used with these catalysts there is some limit or capacity of the catalyst for holding the metals. Apparently, however, this limit is relatively high; it has been suggested that catalysts which have become spent for desulfurization could be used in a guard case to remove arsenic, lead, etc., from the feed to a desulfurizing unit.

XI. Residual Oils and Asphalts

The major portion of the sulfur in a crude petroleum is in the high boiling (above 400°C) fractions and residue. Since burning is the primary use of these materials the sulfur goes into the atmosphere as SO_2. A great deal of effort has been devoted to desulfurizing these materials and commercial processes have been announced (*1*, *14*, *24*), but the commercial catalysts have not been published.

Pilot plant results with a cobalt molybdenum catalyst used at 55 atm pressure, but otherwise within the operating range already given for these catalysts, illustrate the problem of coke deposition (*8*). During the run, sulfur was reduced from 5.1% to 1%. Coke built up rapidly at first but left significant activity. After the initial build up, further coke was deposited more slowly.

Hours on test	37	146	439
% Coke on catalyst	18.0	25.4	29.7

In a different series of fixed bed tests (*6*), it was found that, of the coke deposited in 16 days, one-half was deposited in the first 12 hours. Increasing hydrogen pressure from 70 to 200 atm did not eliminate this initial rapid coke deposition but did lower the "equilibrium" coke level from about 30% by weight of the catalyst to about 18%. (Catalysts containing so much coke are costly to regenerate.)

In a series of tests with active desulfurization components on several different supports, it was found (*6*) that the initial activity was related to the surface area and pore radius in the following way:

$$\% \text{ Desulfurization} = K + 0.0589A + R(6.6A\ 10^{-4} + 0.012)$$

Where: K is a constant depending on reaction conditions, A is the surface area of the catalyst in m^2gm^{-1}, and R is the pore radius in Å. The activity increased with both surface area *and* pore radius. For catalysts with significant internal active surface, the pores must be large enough for the large molecules to enter and to come in contact with the active internal surface.

Systems permitting the introduction and withdrawal of catalysts during operation have been developed as at least a partial solution to the metal contamination and heavy coke deposits (*10*, *14*). Catalyst has been in the form of 1/32 inch extrudates introduced and withdrawn through lock hoppers or in the form of finely divided catalyst introduced suspended in the liquid feed and withdrawn suspended in the liquid product.

In either operation, feed and hydrogen were introduced into the bottom of the reactor. The 1/32 inch catalyst particles were caused to move about (an ebulating catalyst bed) by the feed and internal liquid recycle. With finely divided catalyst, feed and hydrogen provided agitation.

Catalyst consumption was less than 0.1% of oil fed. It was considered small enough so that in some cases it was left in the heavy desulfurized fuel oil. In other cases, most of it was removed by settling and filtering.

Hydrogen consumption was between 1 and 2 moles per mole of feed. This is consistant with the observation that with these large molecules, almost every molecule contains a sulfur atom.

Most of the lower boiling hydrocarbons produced during desulfurization of very heavy oils apparently result from removing an S or N atom that was holding together two or more hydrocarbon groups, rather than from the breaking of C—C bonds.

A thermal treatment of petroleum residues ("visbreaking") before desulfurization seems to decompose some of the less thermally stable components and to reduce coke formation during desulfurization at higher reaction temperatures (*20*).

XII. Poisons

The cobalt molybdenum, nickel molybdenum, and nickel tungsten catalysts are remarkably resistant to poisons. An appreciable amount of CO in the H_2 acts as a poison when treating lower boiling hydrocarbons. Where maximum effectiveness is wanted CO is kept below 50 ppm; otherwise 0.1 to 0.5% can be tolerated.

XIII. Regeneration

The spent catalyst contains metal sulfides as well as the carbonaceous deposits. If iron sulfide has accumulated on top of the bed, it can be pyrophoric and appropriate precautions should be taken. The catalyst is first purged with steam or inert gas. The regeneration is carried out with air diluted with steam or inert gas, so the O_2 content is 0.5–1% and preferably at 2–10 atm pressure. Burning begins at 300°–350°C and the characteristic "hot spot" or burning zone passes through the catalyst bed in the direction of flow. The peak temperature should be kept below 550°C for the cobalt catalysts, and below 450°C for the nickel catalysts. For catalysts containing both Co and Ni, it is safer to use the 450°C maximum.

When the burning zone has passed through the entire catalyst bed, the oxygen concentration can be increased, keeping the above restrictions on the maximum temperature until air itself is being used. Cool to operating temperature or presulfiding temperature and purge with inert gas or steam.

Frequently idle time on a commercial unit is so costly that the deactivated catalyst can be replaced with fresh catalyst much faster and thus more cheaply than regeneration. The used catalyst is then regenerated *ex situ* or custom regenerated. The regenerated catalyst is then ready for the next replacement.

XIV. Presulfiding

Although presulfiding is not necessary in the commercial operation of cobalt molybdenum catalysts, the writer suggests presulfiding for laboratory tests. Presulfiding is necessary with the nickel molybdenum and nickel tungsten catalysts. With catalysts containing both Co and Ni presulfiding is essential if the S content of the feed is low. Again presulfiding is advised for laboratory tests.

The following laboratory procedure is useful. Prepare the reactor at room temperature and pass 10–15% H_2S in H_2 over the catalyst at a rate of 3–4 liters per hour per gram of catalyst. Heat the catalyst to 400°C, taking at least 1.5 hours, while the sulfiding gas flows. Then take the catalyst to operating temperature, stop the H_2S–H_2 flow and proceed with hydrogen pressure and feed for the run.

For sulfiding on a larger scale it should be kept in mind that the sulfiding reaction is exothermic and that a "hot spot" passes through the catalyst very much like a burning zone in a regeneration. Just like in a regeneration, the maximum temperature of the hot spot must be controlled. Heat the catalyst to 315°C and pass 10% H_2S in H_2 over the catalyst at atmospheric pressure and a GHSV of 650 $V_gV_c^{-1}$ hr^{-1}. After the hot spot has passed through the bed, keep at 315°C for 30 minutes while continuing the H_2S–H_2 flow. Heat to 370°C over a period of 1 hour maintaining the gas flow. Cool to operating temperature or slightly below, pressure with hydrogen and begin feed.

If a higher pressure is used during presulfiding, e.g., 15–35 atm, even lower sulfiding temperatures are advised, e.g., 150°–200°C, with a maximum of 240°–250°C. Preferably the catalyst does not "see" either hot hydrogen alone or hot H_2S alone or even high concentrations of H_2S for damaging reactions can occur. After the reaction zone has passed through the bed and the H_2S concentration in the exit gas approaches that in the inlet, the H_2S concentration in the gas is reduced to 5% and the temperature is raised to 300°–310°C at the rate of 50°C per hour and gas flow continued at least 30 minutes after the H_2S content of the exit gas rises towards 5%. Purge with H_2, adjust to reaction conditions, and start feed.

In commercial units, CS_2 has often been found to be a more convenient and more economical source of S than H_2S. The S content of the H_2–CS_2 mixture is kept at about 10% S. The CS_2 may be pumped as such or mixed with recycled hydrocarbons.

Incomplete sulfiding produces catalysts with inferior performance and life. A careful record of sulfur deposited on the catalyst should be kept, and whether H_2S or CS_2 is used, the sulfur "consumed" should closely approach that required to convert NiO into Ni_3S_2, MoO_3 into MoS_2, or WO_3 into WS_2 (7).

With nickel molybdenum catalysts a steam–air pretreatment on fresh or regenerated catalyst (prior to sulfiding) may enhance activity for naphtha and light distillate treating. With the lowest temperature in the catalyst bed at 250°C and the highest temperature 400°C in the presence of air, introduce hot steam with a steam:air weight ratio of 1:1 (as a minimum, 25 volume % air) and heat to 400°C. When the coldest part of the bed reaches 375°C, adjust the steam WHSV to 0.4 $W_l W_c^{-1}$ hr^{-1} and the air concentration to 2.5 volume %. With the entire bed at 400°–425°C, hold the flow and temperature for 16 hours. Maintain the flow and cool the catalyst to 200°–225°C, and adjust to sulfiding temperature in air (7).

XV. Suppliers of Proven Catalysts

American Cyanamid Company

- Aero HDS–2: 3% CoO + 15% MoO_3 (0.025% Na_2O) on alumina, as 1/16 and 1/8 inch particles. The ABD is 0.55, the surface area 275 m^2gm^{-1}, and the pore volume 0.75 cm^3gm^{-1}.
- Aero HDS–6: 3% CoO + 10% MoO_3 (0.025% Na_2O) on alumina, as 1/16 and 1/8 inch particles. The ABD is 0.55, the surface area 275 m^2gm^{-1}, and the pore volume 0.65 cm^3gm^{-1}.
- Aero HDS–3: 3.1% NiO + 15% MoO_3 (0.02% Na_2O) on alumina, as 1/16 and 1/8 inch particles. The ABD is 0.60, the surface area 180–200 m^2gm^{-1}, and the pore volume 0.60 cm^3gm^{-1}. Must be presulfided. Maximum sulfiding temperature should not exceed 200°C.
- Aero HDS–9 and HDS 9A: 3.0–4.0% NiO + 17.5–18.5% MoO_3 (0.04% Na_2O maximum) on alumina as 0.10–0.15 inch (HDS–9) or 0.05–0.075 inch (HDS–9A) diameter extrudates. The ABD is 0.7–0.8. Should be presteamed and must be presulfided. Maximum presulfiding temperature should not exceed 200°C.

Catalysts and Chemicals, Inc.

- C20–6 oxides of cobalt and molybdenum on low sodium, activated alumina, as 1/16, 1/12, and 1/8 inch extrudates. Presulfiding recommended for optimum activity.
- C20–7: oxides of nickel and molybdenum on low sodium alumina, as 1/16, 1/12, and 1/8 inch extrudates. Presulfiding is recommended.

Catalysts Division of Chemetron Corporation

- Girdler G–35: a tabletted 3.5% CoO, 10% MoO_3 catalyst.
- Girdler G–51: an extruded 3.5% CoO, 10% MoO_3 catalyst.
- Girdler G–76: an extruded 2.5% CoO, 2.5% NiO, 10% MoO_3 catalyst.
- Girdler G–85: a spherical 2.0% CoO, 10% MoO_3 catalyst.

Filtrol Corporation

Grade 475–8: 2.8% CoO, 15% MoO_3 as 1/8 inch extrudates. The ABD is 0.75.

Grade GC 4–8, GC 4–12, GC 4–16*: 1.25% Co, 2.25% Ni, 11% Mo, as 1/8, 1/12, and 1/16 inch extrudates. The ABD is 0.8.

Grade GC 3–8, GC 3–16,* 1.85% Co, 0.35% Ni, 10% Mo, as 1/8 and 1/16 inch extrudates. The ABD is 0.8. Recommended for units without an H_2S scrubber in the recycle gas circuit.

Grade GC 25–8, GC 25–16*: 1.25% Co, 2.25% Ni, 11% Mo, as 1/8 and 1/16 inch extrudates. The ABD is 0.8. Recommended for naphtha pretreating under mild conditions to remove both nitrogen and sulfur to less than 1.0 ppm each.

Harshaw Chemical Company

CoMo–0401 T: a silica promoted catalyst containing 3% CoO and 9% MoO_3 on alumina, as 1/8 inch tablets. The ABD is 1.0, the surface area 160 m^2gm^{-1}, and the pore volume 0.40 cm^3gm^{-1}.

CoMo–0402 T: a silica promoted catalyst containing 3% CoO and 15% MoO_3 on alumina, as 1/8 inch tablets. The ABD is 1.0, the surface area 200 m^2gm^{-1}, and the pore volume 0.40 cm^3gm^{-1}.

CoMo–0601 T: a nonsilicated catalyst containing 3% CoO and 9% MoO_3 on alumina as 1/8 inch tablets. The ABD is 1.1, the surface area 160 m^2gm^{-1} and the pore volume 0.40 cm^3gm^{-1}.

CoMo–0603 T: a nonsilicated catalyst containing 3% CoO and 12% MoO_3 on alumina, as 1/8 inch tablets.

HT–200: 3% CoO + 12% MoO_3 on alumina as 1/32, 1/16, 1/12, and 1/8 inch extrudates. The surface area is 180 m^2gm^{-1} and the pore volume 0.42 cm^3gm^{-1}.

HT–100: a silica promoted (1.5%) catalyst containing 3.8% NiO and 16.8% MoO_3 on low sodium (0.02% Na_2O) alumina, as 1/16, 1/12, and 1/8 inch extrudates. The ABD is 0.7, the surface area 175 m^2gm^{-1}, and the pore volume 0.54 cm^3gm^{-1}.

Ni–4303 E: 6% Ni + 19% W as oxides on alumina, as 1/12 inch extrudates. The ABD is 0.8, the surface area 152 m^2gm^{-1}, and the pore volume 0.54 cm^3gm^{-1}. Sulfiding is essential.

Houdry Process and Chemical Company

Series C: 3.2% CoO + 14.6% MoO_3 on low sodium (0.03% Na_2O) alumina, as 1/8 and 5/32 inch extrudates. The ABD is 0.75 and the surface area 330 m^2gm^{-1}.

* Produced under Gulf Research and Development Company license.

Katalco Corporation

41–4: cobalt oxide–molybdenum oxide on alumina. The ABD is 0.6, the surface area 260 m^2gm^{-1}, and the pore volume 0.46 cm^3gm^{-1}.

Nalco Chemical Company

Nalcomo 471: 3.5% CoO + 12.5% MoO_3 on alumina (0.05% Na_2O), as 1/16 and 1/8 inch extrudates. The ABD is 0.64, the surface area 270 m^2gm^{-1}, and the pore volume 0.48 cm^3gm^{-1}.

Nalcomo 474: 3.5% CoO + 12.5% MoO_3 on alumina (0.08% Na_2O) as 1/16 and 1/8 inch extrudates. The ABD is 0.73, the surface area 270 m^2gm^{-1}, and the pore volume 0.51 cm^3gm^{-1}.

NM–501: 5.5% NiO + 13.0% MoO_3 on alumina (0.05% Na_2O), as 1/16 and 1/8 inch extrudates. The ABD is 0.56, the surface area 360 m^2gm^{-1}, and the pore volume 0.50 cm^3gm^{-1}.

NM–502: 3.0% NiO + 14% MoO_3 (0.08% Na_2O) on alumina, as 1/16 and 1/8 inch extrudates. The ABD is 0.73, the surface area 275 m^2gm^{-1}, and the pore volume 0.51 cm^3gm^{-1}.

NT: 5% NiO + 20% WO_3 (0.05% Na_2O) on alumina, as 1/16 and 1/8 inch extrudates. The ABD is 0.80, the surface area 215 m^2gm^{-1}, and the pore volume 0.38 cm^3gm^{-1}.

Union Oil Company

N–21: a proprietary nickel molybdenum catalyst, as 1/16 inch extrudates.

Universal Oil Products Co.

Hydrobon S–3: cobalt molybdenum on alumina, as 1/8 inch pellets. The ABD is 1.

Hydrobon S–6: cobalt, nickel, molybdenum on alumina, as 1/16 inch spheres. The ABD is 0.5.

Hydrobon S–7: cobalt, nickel, molybdenum on silica–alumina, as 1/16 inch spheres. The ABD is 0.6.

REFERENCES

1. Anonymous, *Oil Gas J.* **67,** No. 16, 54 (1969).
2. Arey, W. F., Jr., and Kronenberger, L., *Oil Gas J.* **67,** No. 20, 131 (1969).
3. Bachman, W. A., and Stormont, D. H., *Oil Gas J.* **65,** No. 43, 69 (1967).
4. Bailey, G. W., and Nager, M., *Proc. 7th World Petrol. Congr., Mexico City* **4,** 185 (1967).
5. Benson, D. B., and Berg, L., *Chem. Eng. Progr.* **62,** No. 8, 61 (1966).
6. Beuther, H., and Schmid, B. K., *Proc. 6th World Petrol. Congr., Frankfurt, Main* **3,** 297 (1963).
7. Brewer, M. B., and Cheavens, T. H., *Hydrocarbon Process.* **45,** No. 4, 203 (1966).
8. Gwin, G. T., Heinrich, R. L., Hoffmann, E. J., Manne, R. S., Meyer, H. W. H., Miller, J. R., and Thorpe, C. L., *Ind. Eng. Chem.* **49,** 668 (1957).

9. Hansford, R. C., Gaudio, D. A., Inwood, T. V., and Mavity, V. T., *Oil Gas J.* **67,** No. 18, 134 (1969).
10. Hellwig, L. R., Schuman, S. C., Slyngstad, C. E., and Van Driesen, R. P., *Proc. Am. Petrol. Inst. Sect. III* **42,** 218 (1962).
11. Kartzmark, R., and Gilbert, J. B., *Hydrocarbon Process.* **46,** No. 9, 143 (1967).
12. Kindschy, E. O., Marple, S., Jr., Nager, M., Olson, R. J., and Roque, E. F., *Chem. Eng. Progr.* **61,** No. 3, 69 (1965).
13. Larson, O. A., and Beuther, H., *Am. Chem. Soc. Div. Petrol. Chem. Preprints* **11,** No. 2, B-95 (1966).
14. McFatter, W., Meaux, E., Mounce, W., and Van Driesen, R. P., *Oil Gas J.* **67,** No. 27, 119 (1969).
15. Menzl, R. L., and Webb, W. L., *Hydrocarbon Process. Petrol. Refiner* **44,** No. 5, 202 (1965).
16. Montgomery, D. P., *Ind. Eng. Chem. Prod. Res. Develop.* **7,** 274 (1968).
17. National Petroleum Refiners Association Panel, *Oil Gas J.* **66,** No. 13, 114 (1968).
18. Otwell, G. N., *Oil Gas J.* **66,** No. 46, 78 (1968).
19. Qader, S. A., Wiser, W. H., and Hill, G. R., *Ind. Eng. Chem. Process Design Develop.* **7,** 390 (1968).
20. Schmid, B. K., and Beuther, H., *Ind. Eng. Chem. Process Design Develop.* **6,** 207 (1967).
21. Stormont, D. H., *Oil Gas J.* **66,** No. 8, 40 (1968).
22. Tregilgas, E. T., and Crowley, D. M., *Hydrocarbon Process.* **48,** No. 5, 120 (1969).
23. White, P. J., Jones, J. F., and Eddinger, R. T., *Hydrocarbon Process.* **47,** No. 12, 97 (1968).
24. Yamato, M., and Watkins, C. H., *Hydrocarbon Process.* **47,** No. 5, 131 (1968).

CHAPTER 17

HYDROCRACKING OF PETROLEUM AND COAL

I. Petroleum Distillates

In a way, hydrocracking started in 1927 with the first Bergius plant for commercially hydrogenating brown coal at Luena, Germany. This evolved into supplementary processes for hydrocracking vaporizable middle oils from coal tar in Germany and England and, a little later, into Esso's hydrocracking of petroleum distillates in the United States. Since the early technology aids in understanding current processes, a brief historical review is in order.

It is clear that the most successful early hydrocracking catalyst was pelleted tungsten sulfide (*7, 18*). Typical operating conditions were 250 atm pressure, 400°C with 1700–2700 volumes (STP) of hydrogen circulated per volume of oil feed. The reaction was highly exothermic, so the catalyst was placed in multiple beds in the reactors and cold hydrogen injected between the beds to control the temperature.

The catalyst was rugged and gave hydrocracking even in the presence of high concentrations of nitrogen, sulfur, and oxygen compounds. However, it was such a good hydrogenation catalyst that even with highly aromatic feeds, the gasoline produced was paraffinic and cycloparaffinic and so had relatively poor antiknock quality (catalytic reforming was not then available to improve the antiknock quality).

By 1935, I. G. Farbenindustrie A.-G. had developed a tungsten sulfide on montmorillonite catalyst that partly avoided this oversaturation of the gasoline. However, the catalyst rapidly lost activity in the presence of organic nitrogen compounds so that coal tar oils required pretreatment, although the catayst was suitable for some petroleum stocks without pretreatment.

TABLE XIV

HYDROCRACKING OF QUIRIQUIRE KEROSINE TO MAKE AVIATION GASOLINE

Pressure (atm)	200
Temperature (°C)	380
Recycle gas ($V_g V_l^{-1}$)	1450
LHSV ($V_l V_c^{-1}$ hr^{-1})	1.3
Fresh feed (volume % of total feed)	60
H_2 consumption (vol. per vol. of fresh feed)	390
Aviation gasoline (volume % of fresh feed)	90

In 1937, Esso, in a pilot plant was using the tungsten sulfide on HF treated Terrana clay (a German montmorillonite) catalyst to produce aviation gasoline by hydrocracking. Table XIV gives the conditions and yields (*24*).

The desire for good antiknock quality gasoline led the European coal tar hydrogenators to develop two-stage operations. In the first stage, the old reliable pelleted tungsten sulfide was used, but temperature and space velocity were adjusted so that the prime reaction was nitrogen removal and the hydrogenation of aromatics minimized. Thus, the first stage was a "hydrorefiner" for denitrogenation. Nitrogen contents were reduced to about 5 ppm, while S was reduced to about 0.01%, and oxygen to about 0.1% (7). However, the tungsten sulfide on HF treated montmorillonite catalyst was so sensitive to nitrogen poisoning that sulfuric acid washing was used to further reduce the nitrogen, from 5 ppm to about 1 ppm, before hydrocracking.

In the 1935–1940 period, it became clear that the catalyst was a bifunctional catalyst, i.e., a hydrogenating catalyst and a cracking catalyst. In many respects, the function of the hydrogenating component was to keep the cracking component clean. There also came the realization that while these two functions might be combined in tungsten sulfide, they were separable in the tungsten sulfide on montmorillonite catalysts, so each component could be adjusted for optimum performance. Proper pretreatment of the feed to eliminate poisons permitted a wider choice of catalyst components. By 1939, Imperial Chemical Industries, Ltd. in England had developed an iron on HF treated montmorillonite for the second stage of their two-stage process for hydrocracking middle oils. This catalyst was good enough to form the backbone of Britain's Persuit Grade avaiation gasoline until the end of World War II.

Esso had been seeking better antiknock quality gasolines from hydrocracking petroleum. They found that Ni on HF treated montmorillonite did this, but that the Ni was fairly rapidly poisoned by sulfur compounds. Their two-stage process then was tungsten sulfide on HF treated montmorillonite for the first stage to remove sulfur, and Ni on HF treated montmorillonite for the second stage. The improvement in antiknock quality was relatively small but this was in a critical range. The commercial two-stage unit was operated during 1940–1944 producing aviation gasoline base stock (*24*).

There is a good summary and review giving many details on the hydrocracking of petroleum, so it can be seen how present hydrocracking followed logically from these earlier practices (*25*). The process can be divided into two distinct phases. (1) Removing sulfur compounds which inhibit or poison the hydrogenation component and nitrogen compounds which inhibit the acidic cracking component. (2) Hydrocracking with an acidic cracking catalyst component plus a hydrogenation component that keeps the cracking catalyst clean and hydrogenates unsaturated products to paraffins and cycloparaffins.

A. Phase 1

This is essentially a hydrorefining operation and the reactions, process conditions, and catalysts are those given in Chapter 16. Since the highly aromatic unconverted oils from catalytic cracking are used as raw material for some hydrocrackers, the hydrogenation of these to cycloparaffins is an essential function of Phase 1 along with sulfur and nitrogen removal.

B. Phase 2

In principle, any of the cracking catalysts discussed in Chapter 4 can be used as the cracking component. Table XV gives some comparative results on some cracking components used with several hydrogenating components.

Table XV

Comparative Hydrocracking Catalyst Activities[a]

	Vol. % Conversion to 205°C and lower boiling products[b]
Ni (7%) on HF-treated montmorillonite	68
NiS (7% Ni) on HF-treated montmorillonite	81
NiS (6% Ni) on synthetic SiO_2–Al_2O_3	66
CoO (2% Co) + MoO_3 (7% Mo) on synthetic SiO_2–Al_2O_3	5
CoO (2% Co) + MoO_3 (7% Mo) on HF-treated Al_2O_3	6
Pt (0.2% Pt) on synthetic SiO_2–Al_2O_3	25
Pd (0.5% Pd) on synthetic SiO_2–Al_2O_3	92
Noble metal on "new" support (0.5% metal)	99[c]

[a] See reference (*24*).

[b] At constant LHSV, 75 atm hydrogen pressure, and 315°C. The feed was unconverted material from catalytic cracking which had been hydrorefined to reduce the nitrogen content. 1% CS_2 had been added to the feed for all except the Ni on montmorillonite.

[c] It is the author's opinion that the noble metal was Pd and that the "new" support was the acid form of a molecular sieve.

As this is being written, the commercial hydrocracking catalysts can be said to be in a state of flux, i.e., changes and improvements are frequent. The major hydrogenation components are Pd, NiS_x, or NiS_x–WS_y. The major cracking components are the acid form of a Y-type molecular sieve, synthetic amorphous silica–alumina, or synthetic silica–magnesia. Each process licensor has one or more catalysts that are provided with the license. They are proprietary and the composition is not published.

In commercial units, the second phase will operate at about 350°C, 100 atm pressure, and an LHSV of 0.5–2, with hydrogen recirculation of 350–900 volumes (STP) per volume of liquid. As the catalyst ages, the operating temperature is raised towards 425°C to maintain conversion. With current catalysts this may take 1–3 years. The catalyst may then be replaced or regenerated *in situ*. Conversion to lower boiling products usually is about 60% per pass. Unconverted material may be recycled to extinction.

While the above description of hydrocracking treats it as a two-stage process, it may be a one- or two-stage process, and the operations in the separate stages may be different. With a clean distillate feed, hydrocracking may be conducted in a single stage. As the feed gets dirtier, i.e., contains more sulfur, nitrogen, and aromatic compounds, it is preferable to reduce these in a first stage and to do the hydrocracking in a second stage. The H_2S and NH_3 produced in the first stage may not be as severe catalyst poisons for the second stage as the sulfur and nitrogen compounds originally present. It is thus possible to consider passing the entire product from the first stage to a second reactor without the cost of separating H_2S and NH_3. This technique is applicable to the Pd on a "Y" type deionized molecular sieve.

Even current single-stage units may have two or more reactors in series. Additional hydrogen is added between the reactors, and unconverted material is recycled to the second reactor when additional conversion is desired (*8*, *21*).

The lower boiling fractions (up to 80°–85°C) of the product have acceptable antiknock quality for use in current gasolines but the higher boiling fractions (85°–200°C) usually do not. These are rich in cycloparaffins and are catalytically reformed to produce an acceptable product, while at the same time making some hydrogen for use in the hydrocracking process. Progress is being made in producing acceptable antiknock quality gasoline directly by hydrocracking and eliminating the need for reforming (*8*), even though the current optimum is hydrocracking plus reforming (*13*).

While hydrocracking has, so far, been discussed in terms of making gasoline, it is also used to make jet fuel by altering the fractionation and recycle procedures (*26*). In a different fractionation and recycle operation, higher boiling distillates can be cracked to gases, mostly propane. Highly active Pt on specially prepared chlorided alumina has been studied for hydrocracking low boiling naphtha (C_5–C_8) to gas (*11*).

Broadly, the susceptibility of hydrocarbon types to hydrocracking and the hydrocracking chemistry is similar to that of catalytic cracking (see Chapter 4). Hydrogenatable materials e.g. aromatics and olefins are hydrogenated to saturates before or after cracking. There are a number of differences in the chemistry and the reader interested in the details will want to consult the following references (*22, 23*).

C. Suppliers of Proven Catalysts

Harshaw Chemical Company

Ni–4301 E: 6% Ni + 19% W (as oxides) on silica–alumina, as 1/12 inch extrudates. Must be sulfided. The ABD is 1, the surface area 228 m^2gm^{-1}, and the pore volume 0.37 cm^3gm^{-1}.

Union Oil Company

(To licensees) HC–11: Pd on a "Y" type molecular sieve for use in the first stage of a two-stage hydrocracker or in a single-stage unit.

II. Hydrodealkylation

Catalysts are available for the dealkylation of aromatic hydrocarbons such as the demethylation of toluene to benzene or methylnaphthalene to naphthalene. At first glance, one might consider the catalytic hydrodealkylation of aromatics as hydrocracking, and in a way it is. However, the commercial processes are concerned primarily with removing methyl groups or at most, ethyl groups. The available cracking components of hydrocracking catalysts are not always effective for breaking the bond between the ring and the methyl group, so neither are the hydrocracking catalysts. There is the further complication of the aromatic ring. If the temperature is too low, it is hydrogenated and lost.

Aromatic hydrocarbons can be hydrodealkylated over high purity, low sodium chromia–alumina catalysts in the temperature range of 550°–650°C, pressure 35–80 atm, a LHSV of 1–2 $V_lV_c^{-1}$ hr^{-1}, and with a hydrogen circulation of 8–15 moles per mole of hydrocarbon (*2, 14, 17*). The reaction is strongly exothermic, so careful temperature control is necessary to prevent a runaway reaction.

A. Suppliers of Proven Catalysts

See Chapter 6 for the many suppliers of chromia–alumina catalysts.

III. Demethylation of Isoparaffins

Branched chain paraffins can be demethylated one methyl group at a time by passing them over a prereduced nickel on kieselhuhr catalyst in the temperature range of 200°–320°C, at a pressure of from atmospheric to 50 atm, an LHSV of 0.5 $V_l V_c^{-1}$ hr^{-1}, with a hydrogen:hydrocarbon mole ratio of about 3. It is preferable to adjust the conditions so that the conversion per pass is in the 30–40% range; otherwise, the first reaction product will commence to be demethylated too. By means of this process, it is possible to demethylate isopentane to isobutane, neohexane to neopentane, and 2,2,3-trimethylpentane to 2,2,3-trimethylbutane. At the conversion levels indicated, the process is 90% or better selective for the indicated compound (*12*).

A. Catalyst Suppliers

See Chapter 13 for suppliers of prereduced and stabilized nickel on kieselguhr catalysts.

IV. Hydrocracking for Lubricating Oils

In making lubricating oils for automotive use, oils are preferred which decrease in viscosity as temperature increases as little as possible, i.e., have a low viscosity–temperature coefficient. The types of hydrocarbons wanted for such properties are known, as well as the types not wanted. Interestingly, the unwanted types are readily cracked catalytically to easily separated lower boiling products or they are hydrogenated to intermediates that crack readily. In this way, the desired lubricating oil types are concentrated.

The process begins with vacuum distillates or solvent deasphalted petroleum residues. The aromatics are hydrogenated, at least in part, to cycloparaffins which are then hydrocracked to lower boiling products that are removed, or to desirable lubricating oils in the distillate boiling range. Normal paraffins (wax) are unchanged for the most part and must be removed (*3*). Conditions are 70–200 atm, 350°–400°C, with 1000 volumes of hydrogen circulated per volume of liquid hydrocarbon fed. One catalyst is nickel sulfide–tungsten sulfide on silica–alumina, but there are several process licensors and each has his own preferred catalyst (*1*, *4*, *5*, *15*, *27*).

It is the writer's opinion that the high molecular weight products sought in this process are quite susceptible to further hydrocracking to undesired lighter products. Hence, the activity of the cracking part of the catalyst is critical, i.e., it must not be too active and should be designed with relatively large pores for the large molecules involved. In other words, product yield and product distribution may be diffusion controlled. Also, in the author's opinion, some of the aluminas are acidic enough to supply the cracking function.

A. Catalyst Suppliers

Harshaw Chemical Company

Ni–4301 E: 6% Ni + 19% W (as oxides) on silica–alumina, as 1/12 inch extrudates. Must be sulfided. The ABD is 1, the surface area 228 m^2gm^{-1}, and the pore volume 0.37 cm^3gm^{-1}.

Ni–4303 E: 6% Ni + 19% W (as oxides) on alumina, as 1/12 inch extrudates. Must be sulfided. The ABD is 0.8, the surface area 152 m^2gm^{-1}, and the pore volume 0.54 cm^3gm^{-1}.

Nalco Chemical Company

NT: 5% NiO + 20% WO_3 (0.05% Na_2O) on alumina, as 1/16 and 1/8 inch extrudates. The ABD is 0.80, the surface area 215 m^2gm^{-1}, and the pore volume 0.38 cm^3gm^{-1}.

V. Hydrocracking Petroleum Residuals or Coal

In spite of the vast amount of background information on coal hydrogenation, there is not yet a commercial hydrocracking process for petroleum residuals in which catalyst contamination or coking is not a significant factor. The same problems are present that are discussed in Chapter 16 in connection with the desulfurization of these residuals.

One current process mixes makeup catalyst with the feed and a corresponding amount is removed with the product. The suspension of catalyst and oil is kept in violent motion by means of added and recirculated hydrogen. Thus, the reactor system is a type of fluidized or ebullating bed. The hydrocracking reaction is exothermic but the heat of reaction can be controlled by adding cold feed, cold recycle, cooling the recirculated hydrogen, or all three (*9*, *10*, *20*). Operating conditions are approximately 400°–425°C and 200 atm pressure, with hydrogen circulation about 1000 volumes per volume of oil. With conversion to lower boiling products of 85%, hydrogen consumption is 250–400 volumes per volume of oil. In principle, it is possible to hydrocrack all the oil, but as of this writing, this is not economical.

This same type of unit can be used with coal tar (*20*).

Another method for hydrocracking petroleum residual oils is to deasphalt them first and hydrocrack the deasphalted oils in fixed bed hydrocracking reactors. In this way, the most potent catalyst contaminants do not get to the catalyst (*16*). With fixed bed units, the reactor is designed so that internal quench can be used to control the exothermic reaction.

Hydrocracking catalysts for the processes mentioned above are available to process licensees.

So far as the writer knows, coal is not being hydrogenated and hydrocracked commercially as of this writing. However, because of the recurring interest in the U.S. in motive fuels from coal (*6*) it is pertinent to refer to some of the recently published work. Motive fuels can be made from coal; the only question is at what cost. Nickel sulfide–tungsten sulfide on silica–alumina and cobalt oxide–molybdenum oxide on silica–alumina have been used to hydrocrack deasphalted oil from coal to make 70% gasoline based on the deasphalted oil (*19*).

Molten zinc chloride has been studied in extensive hydrocracking tests on polynuclear hydrocarbons, coal, and coal extracts (*29*). Zinc chloride has an interesting activity for both hydrogenation and cracking especially since it seems to be less sensitive to poisoning by nitrogen compounds than some of the more usual hydrocracking catalysts. It faces some practical problems before it can be commercialized, e.g., catalyst separation from the product, corrosion, and chlorides in the products.

These brief summaries only supplement the wealth of information available on the subject (*28*).

REFERENCES

1. Angulo, J., Gasca, M., Martinez-Cordon, J. L., Torres, R., Billon, A., Derrien, M., and Parc, G., *Hydrocarbon Process.* **47,** No. 6, 111 (1968).
2. Asselin, G. F., and Erickson, R. A., *Chem. Eng. Progr.* **58,** No. 4, 47 (1962).
3. Beuther, H., Donaldson, R. E., and Henke, A. M., *Ind. Eng. Chem. Prod. Res. Develop.* **3,** 174 (1964).
4. Billon, A., Derrien, M., and Levrague, L. C., *Oil Gas J.* **67,** No. 24, 103 (1969).
5. Bryson, M. C., Horne, W. A., and Staugger, H. C., *Oil Gas J.* **67,** No. 23, 90 (1969).
6. Cardello, R. A., and Sprow, F. B., *Chem. Eng. Progr.* **65,** No. 2, 63 (1969).
7. Clough, H., *Ind. Eng. Chem.* **49,** 673 (1957).
8. Duir, J. H., *Hydrocarbon Process.* **46,** No. 9, 127 (1967).
9. Ewing, R. C., *Oil Gas J.* **64,** No. 27, 101 (1966).
10. Ewing, R. C., *Oil Gas J.* **67,** No. 19, 213 (1969).
11. Giannetti, J. P., and Sebulsky, R. T., *Am. Chem. Soc. Div. Petrol. Chem. Preprints* **14,** No. 3, B-36 (1969).
12. Haensel, V., and Ipatieff, V. N., *J. Am. Chem. Soc.* **68,** 345 (1946).
13. Kittrell, J. R., Langlois, G. E., and Scott, J. W., *Hydrocarbon Process.* **48,** No. 5, 116 (1969).
14. McKetta, J. J., Jr., *Advan. Petrol. Chem. Refining* **9,** 46 (1964); **10,** 262 (1965).
15. Morrison, J., *Oil Gas J.* **66,** No. 27, 98 (1968).
16. National Petroleum Refiners Association Panel, *Oil Gas J.* **64,** No. 31, 84 (1966).

17. Natori, B., and De Malde, M., *Hydrocarbon Process. Petrol. Refiner* **43,** No. 12, 91 (1964).
18. Pier, M., *Z. Elecktrochem.* **53,** 291 (1949).
19. Qader, S. A., and Hill, G. R., *Hydrocarbon Process.* **48,** No. 3, 141 (1969).
20. Rapp, L. M., and Van Driesen, R. P., *Hydrocarbon Process. Petrol. Refiner* **44,** No. 12, 103 (1965).
21. Reichle, A. D., and Weller, N. O., *Oil Gas J.* **66,** No. 26, 79 (1968).
22. Sullivan, R. F., Egan, C. J., Langlois, G. E., and Sieg, R. P., *J. Am. Chem. Soc.* **83,** 1156 (1961).
23. Sullivan, R. F., Egan, C. J., and Langlois, G. E., *J. Catalysis* **3,** 183 (1964).
24. Voorhies, A., Jr., and Smith, W. M., *Am. Chem. Soc. Div. Petrol. Chem. Preprints* **7,** No. 1, 187 (1962); *Oil Gas J.* **60,** No. 13, 184 (1962).
25. Voorhies, A., Jr., and Smith, W. M., *Advan. Petrol. Chem. Refining* **8,** 168 (1964).
26. Watkins, C. H., and Jacobs, W. L., *Oil Gas J.* **67,** No. 24, 74 (1969).
27. Watkins, C. H., and Webb, T. A., *Oil Gas J.* **67,** No. 26, 112 (1969).
28. Wu, W. R. K., and Storch, H. H., Hydrogenation of Coal and Tar. *U.S. Bur. Mines, Bull.* **633** (1968).
29. Zielke, C. W., Struck, R. T., Evans, J. M., Costanza, C. P., and Gorin, E., *Ind. Eng. Chem. Process Design Develop.* **5,** 151, 158 (1966).

CHAPTER 18

OXIDATION (INORGANIC)

I. Sulfur Dioxide to Sulfur Trioxide

In terms of the tons per day of material produced, the oxidation of sulfur dioxide to sulfur trioxide is undoubtedly the most important vapor phase catalytic oxidation process. It dates back to 1831, when Phillips showed that platinum was capable of catalyzing the reaction. This was even before Berzelius coined the terms "catalyst" and "catalysis." In 1898, Hasenbach and Clemm described a commercial process using ferric oxide as the catalyst. The next year, R. Meyers indicated that vanadium pentoxide was a good catalyst.

Today, so far as the writer knows, only platinum based and vanadium pentoxide based catalysts are used commercially. As might be expected, there are several variations on the platinum and vanadium catalysts.

A. Platinum Catalysts

One process uses a platinized asbestos containing 8–10% Pt and this is placed in layers ½–¾ inch deep in numerous beds on preforated trays. A second process uses platinized magnesium sulfate prepared by impregnating calcined granules of magnesium sulfate with an aqueous solution of platinic chloride so that the catalyst contains 0.1–0.3% Pt.

1. *Poisons*

The platinum catalysts are poisoned by chlorine compounds and by arsenic.

B. Vanadium Catalysts

At least four somewhat different vanadium catalysts have been used. One is prepared by impregnating kieselguhr with ammonium metavanadate and potassium hydroxide, after which the catalyst is pelleted and calcined in the presence of SO_2 and air. The KOH is converted to potassium sulfate or pyrosulfate during the treatment.

A second catalyst is formed from a solution of ammonium metavanadate and potassium silicate by reacting with hydrochloric acid to form the silica–vanadia gel.

A third catalyst is formed from a base prepared from potassium silicate and kieselguhr. This slurry is reacted with potassium aluminate to form a gellike synthetic "zeolite." This "zeolite" is then impregnated with the vanadium compound formed by the addition of ammonium vanadate to a solution of potassium aluminate and this mixed with a solution of potassium silicate, where again a "zeolite" mass is formed. The finished product is pelleted and calcined in a stream of SO_2 and air.

A fourth catalyst is prepared by dissolving vanadium pentoxide in sodium hydroxide and adding potassium sulfate, gum tragacanth, and kieselguhr. The mixture is neutralized with sulfuric acid, dried, pelleted, and calcined.

Since the commercial catalyst manufacturers do not state how their catalysts are made, this information is given as background (*6*, *9*). It is fairly certain that under operating conditions, the vanadium pentoxide alkali sulfate–pyrosulfate catalysts consist of a highly viscous melt which covers the carrier as a thin film.

The SO_2, O_2—SO_3 system is an equilibrium system with SO_3 favored by relatively low temperatures and $SO_2 + O_2$ by high temperatures. Employing adiabatic reactors, it is necessary to control the maximum temperature to obtain 95+% conversion of the SO_2. This is done by having the catalyst in several separate beds with arrangements for cooling the gaseous reactants between beds.

Most often the SO_2 is obtained by burning S in air so that the mixture going to the catalytic converter contains 6–10% SO_2. With the vanadium catalysts, the beds are usually in one large diameter reactor and are relatively thin to minimize pressure drop. The total pressure is very little over that required to force the gases through the system. The inlet temperature is 400°–425°C, and the outlet of the first bed is 600°C maximum. The interstage coolers drop the temperature to 400°–425°C. The temperature rise in succeeding beds is progressively less so that the rise in the last bed is only 20°–25°C with a final SO_2 conversion of 95+%. The total GHSV is 500–1000 $V_gV_c^{-1}$ hr^{-1} (SO_2 GHSV 50–100 $V_gV_c^{-1}$ hr^{-1}). Depending on the ABD of the catalyst, 100–200 liters of catalyst are required per ton-day of 100% H_2SO_4 production.

1. *Poisons*

Chlorides and Se are temporary poisons for the V catalysts.

C. Suppliers of Proven Catalysts

American Cyanamid Company supplies Aero SA which is a vanadium pentoxide catalyst.

Catalyst and Chemicals, Inc.
C101–2, 9–10% V_2O_5 + 14–15% "K_2O" supported on kieselguhr, as 3/16 extrudates or $\frac{1}{4} \times \frac{1}{8}$, $\frac{1}{4} \times \frac{1}{4}$, and $\frac{3}{8} \times \frac{1}{4}$ inch tablets. The ABD is about 0.6.

Engelhard Industries as well as Matthey Bishop, Inc. supply platinum catalysts.

II. H_2S to Sulfur

Many natural gases and many petroleum refinery gases contain H_2S which must be removed. Depending on the price of sulfur, when the quantity of H_2S exceeds about 10 tons per day, it becomes economical to recover the sulfur for sale. In 1968, the U.S. capacity for converting H_2S to sulfur was over 3×10^6 tons per year and Canadian capacity was over 5×10^6 tons per year (*11*).

In 1885, Chance and Claus described a commercial catalytic process in which H_2S was reacted with air in the presence of an iron oxide catalyst at about 250°C to produce sulfur which was collected as a liquid.

The modern versions of the process take several forms, e.g., partial combustion, split stream, and direct oxidation.

In the partial combustion process (used with gases containing 50–100% H_2S), the H_2S is burned noncatalytically with air which is limited to that required for the reaction:

$$H_2S + \tfrac{1}{2}O_2 = S + H_2O$$

A major part of the H_2S is converted into sulfur that is recovered as a liquid by cooling the reaction products to about 190°C. A part of the H_2S is left unconverted and a part is oxidized to SO_2. The mixture is preheated and fed to the first catalytic reactor, where the following reaction takes place:

$$2\,H_2S + SO_2 = 3\,S + 2\,H_2O$$

The reaction is exothermic and the temperature rises to about 330°–340°C. Again the reaction mixture is cooled to about 190°C to recover the sulfur. The remaining gases still containing H_2S and SO_2 are preheated and fed to a second catalytic reactor where more reaction occurs and the temperature rises to around 275°C. The sulfur formed is recovered by cooling (*10, 11, 14, 15*). The sulfur recovery depends on the initial H_2S concentration but usually exceeds 90%.

In the split stream process (used with streams containing 15–50% H_2S), the gas stream is divided into two parts. One part, one-third of the H_2S, is burned noncatalytically to SO_2. This is mixed with the other two thirds of the H_2S and the mixture sent to the catalytic reactors, where the following reaction occurs, just as it did in the partial combustion process.

$$2\,H_2S + SO_2 = 3\,S + 2\,H_2O$$

In fact, the catalytic part of the two processes is essentially identical (*10, 11, 14, 15*).

The direct oxidation process is applied to gases too dilute in H_2S to support combustion in a noncatalytic burner (2–15% H_2S). The gas is preheated, mixed with air, and fed directly to the catalytic reactor, where H_2S is converted into S and SO_2. More than one reactor can be used with interstage cooling for sulfur separation just as was done in the above processes. More air is introduced between stages (*2, 10, 11, 14, 15*).

In all of these processes bauxite is the catalyst. The pressure is just high enough to force the materials through the plant. Catalyst beds are usually 1 m or less thick to hold down pressure drop and catalyst particles are large for the same reason. Linear velocities through the bed are 0.15–0.6 m sec^{-1}, and the GHSV is in the range of 1000–5000 $V_g V_c^{-1}$ hr^{-1}. Depending on usage, catalyst life is 1–5 years.

A. Supplier of Proven Catalysts

Minerals & Chemicals Division of Engelhard Minerals & Chemicals Corp.
  Porocel: a low-iron, activated abuxite, as 2–4 and 4–8 mesh particles.
  Porocel LDP: a pelleted, activated bauxite.

III. Ammonia to Nitric Acid

In 1839 Kuhlman introduced a commercial process for oxidizing ammonia to oxides of nitrogen which were dissolved in water to form nitric acid. In 1903, Ostwald extended the process.

For discussion, the processes can be divided into "high" pressure, "medium" pressure and "low" pressure. The high pressure units operate at about 8 atm pressure and 900–950°C. The medium pressure units operate at around 4 atm and 850°C–880°C, while the atmospheric pressure units operate at about 800°C. In all processes, a mixture of air and NH_3 containing 9.5–10.5% NH_3 is preheated and passed through 10–30 layers of 90% Pt–10% Rh 80–100 mesh gauze sheets stacked one directly on top of the other. The gauze is made of wire 0.04–0.125 mm in diameter. The reaction is extremely fast, with residence times of the order of 10^{-3} to 10^{-4} seconds, and a GHSV exceeding 10^6 $V_g V_c^{-1}$ hr^{-1}. The physical transport of NH_3 to the catalyst surface is the rate determining step. The reaction is:

$$4\,NH_3 + 5\,O_2 = 4\,NO + 6\,H_2O$$

If the gas flow rate is too high, some ammonia passes through and is lost via the reaction:

$$4\,NH_3 + 6\,NO = 5\,N_2 + 6\,H_2O$$

Too low a gas rate can lose NO by the reaction:

$$2\,NO = N_2 + O_2$$

About 80–90% of the reaction is completed in the top 20% of the gauze layers.

Since ammonia and air form an explosive mixture containing 12.4% NH_3 at 8 atm, the NH_3 concentration is kept well on the "lean" side; hence the 9.5–10.5% NH_3 concentration. Both air and ammonia are filtered to keep out dust or oil before preheating, and are preheated separately. Since iron oxide is a potent catalyst for decomposing NH_3, the preheaters are made of stainless steel or other alloys. Even so, it is estimated that 1.5–3% of the ammonia decomposes before it reaches the catalyst.

The preheated air and ammonia are thoroughly mixed to be sure a rich mixture will not cause a "hot spot" on the catalyst gauze and damage it.

From 25 to 50 kilograms of NH_3 are converted per day per ounce of catalyst at atmospheric pressure. This is increased to about 180 kg at 8 atm. This has its cost, for at 8 atm and 900°–950°C, catalyst losses are 250–400 mg per ton of 100% HNO_3. The "medium" pressure process losses are lower, 180–200 mg per ton, and the atmospheric process still lower, 50 mg per ton (*5*). Some of this catalyst is recovered in a downstream filter.

The hot gases leaving the catalyst are cooled, the NO oxidized (non-catalytically) to NO_2 which is absorbed in water to yield nitric acid (*1, 7*). The residual gas always contains some NO and this is reduced as described in Chapter 12.

Much of the nitric acid made is 50–70% HNO_3. A process for 100% HNO_3 first liquifies the NO_2. It is then converted to HNO_3 in an autoclave with O_2 and H_2O (*12*).

A. Suppliers of Proven Catalysts

Both Engelhard Industries and Matthey Bishop, Inc. supply the Pt–Rh gauze catalysts.

IV. Ammonia to Nitrous Oxide

In 1914, Badische Anilin u. Sodafabrik announced a commercial process for oxidizing ammonia to nitrous oxide using a Bi_2O_3–Fe_2O_3 catalyst. It is of considerable historic interest in that Bi compounds have not been commonly used in catalysts, and it is only within the last 10 years that they have reappeared in oxidation catalysts.

Operating conditions are 300°–400°C and atmospheric pressure, with 80% yields of N_2O. Higher temperatures favor the formation of higher oxides of nitrogen. Suitable catalysts are

80% Fe_2O_3	12% Bi_2O_3	8% MnO_2
36% CuO	64% MnO_2	
45% CuO	45% MnO_2	10% $CaCO_3$

The resemblance of the copper oxide–manganese oxide catalysts to the hopcalite catalysts is obvious. The author does not know of any commercial suppliers of these specific catalysts (*8*, *13*, *16*, *17*).

V. HCl to Chlorine

This is the Deacon process, and it is probably the first industrial fixed bed catalytic process. In this process, HCl and air are passed over a supported $CuCl_2$ catalyst at 430°–500°C to produce chlorine and water. At temperatures above about 500°C, the $CuCl_2$ begins to melt and become sticky. Below 430°C, the rate is too slow to be interesting. Sulfur and arsenic compounds are catalyst poisons. The reaction is an equilibrium reaction:

$$2\,HCl + O_2 \rightleftharpoons Cl_2 + H_2O$$

It is worth noting that there is a commercial version of the process using an undisclosed catalyst in a fluidized solids reactor, with air or oxygen as the oxidizing agent (*3*).

In a different version, nitrogen oxides are used as a homogeneous catalyst replacing the copper chloride and the equilibrium is displaced towards Cl_2

by absorbing the water coproduct in sulfuric acid (*4*). Oxygen is used as the oxidizer.

A. Catalyst Supplier

Harshaw Chemical Company

Cu–0905 T: 10% $CuCl_2$ mounted on activated alumina, as ⅛ inch tablets. The ABD is 1, the surface area 59 m^2gm^{-1}, and the pore volume 0.26 cm^3gm^{-1}.

Cu–0905 P: 10% $CuCl_2$ on activated alumina, as powder. The ABD is 1 and the surface area 180 m^2gm^{-1}.

Cu–0804 T: 7.7% $CuCl_2$ on activated alumina, as ¼ inch tablets. The ABD is 0.9, the surface area 100 m^2gm^{-1}, and the pore volume 0.36 cm^3gm^{-1}.

REFERENCES

1. Anonymous, *Hydrocarbon Process.* **44,** No. 11, 246 (1965); **46,** No. 11, 207 (1967).
2. Anonymous, *Hydrocarbon Process.* **46,** No. 11, 225 (1967).
3. Anonymous, *Hydrocarbon Process.* **46,** No. 11, 166 (1967).
4. Anonymous, *Chem. Eng. News* **47,** No. 17, 14 (1969); **47,** No. 19, 14 (1969).
5. Connor, H., *Platinum Metals Rev.* **11,** No. 1, 2 (1967).
6. Dixon, J. K., and Longfield, J. E., *Catalysis* **7,** 322 (1960).
7. Dixon, J. K., and Longfield, J. E., see reference 6, p. 281.
8. Dixon, J. K., and Longfield, J. E., see reference 6, p. 294.
9. Duecker, W., and West, J. R., "Sulfuric Acid." Reinhold, New York, 1959.
10. Goar, B. G., *Hydrocarbon Process.* **47,** No. 9, 248 (1968).
11. Grekel, H., Palm, J. W., and Kilmer, J. W., *Oil Gas J.* **66,** No. 44, 88 (1968).
12. Hellmer, L., *Hydrocarbon Process.* **45,** No. 11, 183 (1966).
13. Kobe, K. A., and Hosman, P. D., *Ind. Eng. Chem.* **40,** 397 (1948).
14. Opekur, P. C., and Goar, B. G., *Hydrocarbon Process.* **45,** No. 6, 181 (1966).
15. Valdes, A. R., *Hydrocarbon Process. Petrol. Refiner* **43,** No. 3, 104 (1964); **43,** No. 4, 122 (1964).
16. von Nagel, A., *Z. Elektrochem.* **3,** 754 (1930).
17. Zawadzki, J., *Discussions Faraday Soc.* **8,** 140 (1950).

CHAPTER 19

OXIDATION (ORGANIC) LIQUID PHASE

Many of the catalysts used for oxidizing organic chemicals in the liquid phase are inorganic chemicals rather than specially prepared catalysts. Usually they are based on elements that change valence fairly readily, e.g., Co, Cu, Mn. Often they are combined with a fairly large organic fragment either as a salt or chelate to make them soluble in the organic substance to be oxidized. Some of the oxidations have free radical characteristics and the addition of the bromide ion can promote the reaction presumably via the bromine atom as a chain carrier.

I. Ethylene to Acetaldehyde

This is the Wacker process in which ethylene is oxidized to acetaldehyde by means of an aqueous solution of cupric chloride containing traces of palladium chloride. It can be used in a single-stage process or a two-stage process for making acetaldehyde (*2*). In the reaction, the palladium chloride is first reduced to elemental palladium while oxidizing the ethylene to acetaldehyde and is promptly reoxidized to palladium chloride by the cupric chloride. During catalyst regeneration, the cuprous chloride is reoxidized to cupric by oxygen, the ultimate oxidizing agent (*8*).

A. Single-Stage Process

Ethylene, oxygen, and recycled gas are fed into a vertical reactor where intimate contact is made with the catalyst solution under a slight pressure.

The heat of reaction has to be carried away, and this is done by boiling off some of the water. The gases are scrubbed, and the resulting acetaldehyde solution is separated by fractional distillation. The unreacted gas is recycled to the reactor minus a small bleed stream to eliminate impurities in the feed. In this process, practically pure ethylene and pure oxygen are used.

B. Two-Stage Process

In this operation the ethylene is absorbed by contact with the solution of catalysts, with the net result that the copper is reduced from cupric to cuprous and the ethylene is converted to acetaldehyde. This operation is carried out under pressure, and the solution is withdrawn from the reactor. The pressure is reduced and the acetaldehyde flashed off with some of the water of the solution. The acetaldehyde is recovered by fractional distillation. The used catalyst is taken to a second reactor where it is mixed with air; the oxygen reoxidizes the cuprous chloride to cupric and the catalyst solution is recycled to the reactor.

In the two-stage process, it is feasible to use hydrocarbon streams that are not pure ethylene, e.g., mixtures in petroleum refinery gases, and it is also possible to use air instead of the pure oxygen required in the single-stage process. For these reasons the two stages process has a certain attractiveness to it.

The yield of acetaldehyde is about 95% in either process.

II. Acetone from Propylene and Methyl Ethyl Ketone from Butenes

In this extension of the Wacker process, the two-stage operation as described above for ethylene is used. The olefin under pressure is absorbed and oxidized by the cupric chloride–palladium chloride catalyst solution at 90°–120°C and 9–12 atm pressure. The reacted solution is taken to a second vessel where the pressure is reduced to 1 atm and the ketone stripped from the solution by steam. The catalyst solution is moved to a third vessel where it is reoxidized for reuse by air under pressure.

The rate of reaction of these higher olefins is considerably slower than ethylene. The rate of absorption of pure propylene is about one-third that of ethylene and, in terms of catalyst usage, the acetone production rate is 130 gm hr^{-1} $liter^{-1}$. The rate of reaction of the butenes is still slower with 1-butene reacting about one-seventh as fast as ethylene and 2-butene reacting about one-fourteenth as fast. If the olefins are mixed with paraffins, i.e., commercial streams, the rate is still slower. In terms of catalyst usage, methyl ethyl ketone production is about 30 gm hr^{-1} $liter^{-1}$.

With propylene, 6–8% byproducts such as propionaldehyde, chloroacetone, and 1,1-dichloroacetone are formed, With butenes, butyraldehyde, 3-chlorobutanone-2, and 3,3-dichlorobutanone-2 are formed. In both cases, minor amounts of products are formed containing one or two carbons less that the starting olefin. Needless to say, these byproducts complicate the problem of producing pure ketones. The yield of acetone is 92–94% and of methyl ethyl ketone 85–88% (*9*, *38*).

A. Corrosion

These catalyst solutions are extremely corrosive to ferrous metals. Titanium or other chloride resistant materials are advisable.

B. Catalyst

The catalyst for these three processes is 300–400 mg Pd^{+2} per liter plus a large excess of cupric chloride, e.g., 50–100 gm $liter^{-1}$ in water. Hydrochloric acid is added from time to time to replace that consumed in forming chloro compounds.

III. Vinyl Acetate from Ethylene (Liquid Catalyst)

This process resembles the single-stage acetaldehyde process except the redox catalyst system is dissolved in acetic acid instead of water. Furthermore, some alkali acetate, e.g., sodium acetate, must be present to buffer the system and an alkali chloride, e.g., lithium chloride, to maintain a rather critical Cl^- concentration. (Since some Cl^- is consumed forming chloro byproducts, it must be replaced.) The overall reaction can be represented:

$$C_2H_4 + CH_3COOH + \tfrac{1}{2}\,O_2 = H_2C{=}CH{-}O{-}COCH_3 + H_2O$$

The H_2O present in the catalyst solution permits the formation of acetaldehyde, and the ratio of vinyl acetate to acetaldehyde is controlled by controlling the H_2O content of the catalyst solution (*11*, *30*). The solution contains about 35–40 mg Pd^{+2} and 3–6 gm Cu^{+2} per liter.

The process operates at about 120°C and 30 atm pressure. A mixture of around 5% O_2 + 95% C_2H_4 is intimately mixed with the liquid catalyst, e.g., in a bubble column. (The O_2 concentration is on the rich side, but close to the explosion limit which, for example, is 5.5% O_2 in ethylene at 130°C and 3 atm.) Around one-half of the O_2 reacts per pass, along with a corresponding amount of C_2H_4. The gas is used to strip the reaction products from the catalyst solution and is then quickly chilled, e.g., to 0°C, to separate the liquid products and unreacted acetic acid from the gas. The unreacted gas containing oxygen and ethylene is fortified with more O_2 and C_2H_4 and recycled to the catalyst (*11*, *30*).

If the Pd^{+2} concentration gets too high, e.g., 300 mg Pd^{+2} per liter, ethylene is lost by polymerizing to *n*-butenes. Water inhibits this reaction (*30*).

The yield of vinyl acetate + acetaldehyde + a little acetic formed in the process is 90% based on the ethylene, and 95% based on the acetic acid. The minimum acetaldehyde corresponds to a vinyl acetate:acetaldehyde ratio of about 1:0.2.

IV. Vinyl Acetate from Ethylene (Solid Catalyst)

In this version of the process, a mixture of ethylene, oxygen, and acetic acid is passed over a supported Pd catalyst in a mixed phase or vapor phase operation. Typically, ethylene is passed through heated acetic acid to give the desired ratio and then oxygen is introduced. The mixture is passed over the solid catalyst contained in the tubes of a heat exchanger. There is boiling water, under pressure, outside the tubes to carry away the heat of reaction. The product is cooled rapidly to separate liquid products and unreacted acetic acid. Only part of the oxygen and ethylene react, so after scrubbing to remove CO_2, they are recycled (*30*, *34*, *37*). The yield of vinyl acetate is 90+% based on the ethylene, and the acetaldehyde yield is 1% or less.

More specific conditions are: Temperature 160°C, pressure 5 atm, total GHSV 1000 $V_g V_c^{-1}$ hr^{-1} (made up of, e.g., 62 mole % C_2H_4, 31 mole % acetic acid, and 7 mole % O_2). The conversions per pass are about 60–70% of the O_2, 20% of the acetic acid, and 10% of the ethylene. The production rate is 100–200 gm hr^{-1} $liter^{-1}$. The O_2 content of the circulating gas is about 2.5%, well below and on the rich side of the explosion limit.

The Pd is supported on a material resistant to hot acetic acid, e.g., carbon, some aluminas, or silica, and the Pd content is 0.2–2.0%. The catalyst is modified by 0.5–5% alkali acetate (*30*) which is replaced as it is lost by the catalyst (*37*) or by small amounts of cupric and ferric salts (*34*).

A. Suppliers of Proven Catalysts

The author does not know of suppliers of the specific catalysts used here. The following supply Pd catalysts supported on carbon or alumina. The reader preferring to use an alumina supported catalyst will confirm with the supplier that it is a form insoluble in hot acetic acid.

Calsicat Division of Mallinckrodt Chemical Works supplies Pd on carbon

Davison Chemical Division of W. R. Grace & Co. supplies Grade 942, 0.2% Pd on 4 to 6 mesh carbon granules.

Engelhard Industries or Johnson Matthey, Inc.
Pd on charcoal: granules, or extrudates.
Pd on alumina: granules, tablets, or extrudates.

Harshaw Chemical Co.
Pd–0501 T: 0.3% Pd on high activity alumina, as ⅛ inch tablets. The ABD is 0.9, the surface area 186 m^2gm^{-1}, and the pore volume 0.40 cm^3gm^{-1}.

V. Alkyl Side Chains to Aromatic Acids

Some of the elements with variable valences have the ability to catalyze the oxidation of many substances: the alkyl aromatics to be discussed here, as well as saturated hydrocarbons, unsaturated glycerides (siccatives in paints, etc. The element with widest use is cobalt. Manganese, cerium, and molybdenum are also used. The elements are used as organic compounds so they will be soluble in the substrate.

The cobalt catalysts, at least, have certain interesting characteristics. Starting with Co(II), there is an induction period, presumably, while it is oxidized to Co(III) by oxygen. With alkyl aromatics, the Co(III) in acetic acid can oxidize the hydrocarbon in the absence of added oxygen. Apparently Co(II) actually inhibits this reaction (*23*). Activators such as Br^- (*41*), acetaldehyde (*31*), or methyl ethyl ketone (*15*) are used. These are thought to be regenerable sources of chain-initiating free radicals.

The following laboratory oxidation procedures, although not complete, may be useful.

Five liters of toluene was charged to a reactor equipped for moderate pressure and reflux. The 0.1% of cobalt catalyst, e.g., cobalt benzoate, was dissolved in the toluene and the mixture heated to 130°–140°C. Air was fed in. There was a short induction period after which oxidation proceded smoothly. The benzoic acid concentration was allowed to build up to 25–30%, after which 500–750 ml of solution was removed each hour and replaced with toluene containing 13 mg of fresh cobalt catalyst. The benzoic acid was separated and the toluene recycled (*10*, *27*, *28*).

The following procedure uses acetic acid as a solvent and includes the Br^-. 75 gm of *p*-xylene was dissolved in 150 gm of acetic acid and 0.75 gm of NH_4Br + 1 gm of cobalt acetate tetrahydrate added. The mixture was heated to 195°–205°C (27 atm pressure) and oxidized with air for 2 hours using reflux to control the temperature. The yield of terephthalic acid was 75–80% (*17*, *41*).

Some hydrocarbons, e.g., xylenes, absorb oxygen so readily under these conditions that the reaction may become starved for oxygen. If this happens, undesired products, some colored, are formed.

The hot bromine-containing system is highly corrosive, so special metallurgy is required in commercial operation. This is especially true since iron and copper compounds can inhibit or poison the reaction.

An entirely different method uses NO_2 as the oxidizing agent and Se as the catalyst (*32*). The following laboratory procedure may be useful. Two liters of trichlorobenzene were put in a 3 liter flask (good stirrer) and 200 gm of 2,6-dimethylnaphthalene + 2 gm of Se added. The mixture was heated to 195°C and NO_2 passed in (good stirring) until the brown fumes of unreacted NO_2 appeared in the exit gas. The product was cooled and the 2,6-naphthalene dicarboxylic acid recovered by filtration.

VI. Siccatives (Paint Driers)

Catalysts known as siccatives or "driers" have been added for years to oil based paints and varnishes, as well as alkyd resins, to speed up "drying" or hardening. The parallel of the action of some of the siccatives with the catalysts in the preceding section is an interesting one. The variable valence elements are similar and they are combined with organic molecules, so they will be soluble. The elements Co and Mn at room temperature and Ce at "baking" temperature initiate drying via the formation of oxidative intermediates. Other elements such as lead, zinc, calcium, and zirconium complement the action of the Co and Mn by aiding in the polymerization. Without Co or Mn to initiate drying processes, the total polymerization reaction would proceed more slowly (*40*).

Cobalt (as an oil soluble compound) is used to the extent of 0.005–0.2% (as metal) based on the weight of the oil. It aids drying at the surface so a tack-free film is produced rapidly.

Manganese is ordinarily used with a cocatalytic element such as lead, e.g., 0.02% Mn (as metal) + 0.6% Pb, based on the oil. Mn is especially effective in coatings to be baked where 0.005–0.3% is used.

Lead aids by promoting the polymerization of the oxidative intermediates formed by the primary driers. The amount used varies from 0.1–5% Pb based on the oil. As Co promotes a tack-free surface, Pb causes polymerization throughout the film to form a hard but flexible coating. Lead can be used without Co or Mn, but the drying is much slower.

In the presence of Co or Mn, Ca can be used to replace Pb, partly or completely. One-fourth to one-third the weight of Ca as Pb is needed.

Zirconium behaves somewhat like lead but it cannot be used alone. Used with Co, it, like Ca, can be used to produce lead-free products.

Zinc, used with cobalt, inhibits free radical chain branching and thus prevents wrinkling of the film. A more porous surface is formed so more oxygen can be absorbed by the film before the surface hardens. The amount used is 0.01–1%, i.e., about two to three times the weight of Co present.

Iron acts as a drier but its use is limited to dark finishes because of its color. Its polymerization effect increases at elevated temperatures and this is used to advantage in finishes to be baked. About 0.02–0.1% is used.

The activity of Ce is relatively low at room temperature. It increases with temperature and is good at baking temperature, where it replaces iron in light colored products.

Cu is not a catalyst and it may be a poison (inhibit drying) (*40*). Cu(II) is known to be a free radical intercepter (*29*).

A. Suppliers of Proven Catalysts

Many different soaps, salts, or chelates of these metals are manufactured to make them soluble in organic media. As a result, the metal content can vary rather widely. Co, Mn, and Fe driers average about 6% metal content. The Ca content tends to be somewhat lower, 4–5%, the Zn content higher, 8%, and the Pb content still higher, 16–24%. Because of the many differences, the individual products are not listed but just the suppliers.

Carlisle Chemical Works, Inc.
Harshaw Chemical Company
MacKenzie Chemical Works, Inc.
Nuodex Division of Tenneco Chemicals, Inc.
Witco Chemical Corporation, Organics Division

VII. Aromatic Acids to Phenols—Benzoic Acid to Phenol

The second step in the Dow phenol process involves the oxidation of benzoic acid to phenol in the presence of a copper catalyst (copper benzoate), oxygen, and water. Apparently benzoyl salicylic acid and salicylic acid are intermediates.

The following laboratory procedure may be useful: 250 gm of benzoic acid was melted in a beaker and 13.6 gm of MgO plus enough cupric carbonate to make 0.069 gm moles of copper benzoate was added; the mixture was heated to 210°C. This was poured into the preheated reactor which was equipped with an efficient stirrer. At 230°C, air was fed in at 500 cm^3 min^{-1} and water at 1 gm min^{-1}. At first, subliming benzoic acid carried by the exit gases plugged the exit line to the fractionating column. These plugs were removed mechanically. As soon as some phenol formed, a low melting mixture resulted which caused no trouble. Phenol and water were removed continuously by fractionation while unreacted benzoic acid was recycled and makeup acid added (*27*, *28*).

A. Toluic Acids to Cresols

The toluic acids could be oxidized to cresols in an analogous way. The *o*- and *p*-toluic acids gave pure *m*-cresol, while *m*-toluic acid gave a 40% *ortho*–60% *para* mixture.

VIII. Saturated Hydrocarbons to Alcohols

As one attempts to oxidize *n*-paraffins, most often the chain breaks and lower molecular weight products are obtained, usually as a complex mixture (*25*, *36*). However, the higher *n*-paraffins can be oxidized in the presence of boric oxide to yield a mixture of secondary alkyl borates in which the alkyl group contains the same number of carbon atoms as the paraffin (*12*). Aromatic hydrocarbons inhibit the reaction, even in small concentrations, and so should be removed, e.g., by silica gel (*25*). The reaction can be carried out in a simple stirred reactor at 155°–170°C and atmospheric pressure, or under pressure if need be, with a suspension of boric acid or oxide. Air or air diluted with nitrogen to 2–5% O_2 is passed into the mixture. Conversion is limited to 10–20% before removal of the alkyl borates; otherwise ketone formation increases.

The secondary borates formed are readily hydrolyzed to recover the alcohols.

The branched chain saturates containing a tertiary hydrogen can be oxidized to *t*-hydroperoxides with out a catalyst.

Cyclohexane can be oxidized to cyclohexanol borate (and cyclohexanone) in the presence of boric oxide (*1*, *16*). Conditions for a batch operation are as follows. Temperature 160°–170°C, pressure 12 atm, O_2 in oxidizing gas 4–10%, 30–80 gm of boric acid per kg of cyclohexane, reaction time 90–150 min. The conversion is about 12% and the ratio of cyclyhexanol to cyclohexanone is about 10. The yield of cyclohexanol + cyclohexanone is 85–90% of the cyclohexane converted. A continuous process uses the same principles (*1*).

Cyclododecane can be oxidized to cyclododecanol and cyclododecanone using the same technique (*13*).

IX. Acetaldehyde to Acetic Anhydride

In its simplest form, this reaction can be represented by the equation:

$$2\ CH_3CHO + O_2 = (CH_3CO)_2O + H_2O$$

Since the acetic anhydride and water will react if left in contact, a solvent for the acetic anhydride that keeps them separated is used. The actual reaction mechanism is complex, involving peracetic acid, acetaldehyde monoperacetate, and diacetyl peroxide. The catalysts are cobalt plus copper acetates or manganese plus copper acetates. All three can be used (*18*, *22*).

The reaction is conducted in a stirred reactor at 50°–60°C under pressure. Isopropropyl acetate + acetic acid may be used as the solvent to protect the acetic anhydride. The catalyst is added as a solution of the metal acetates in acetic acid–acetic anhydride solvent. Reaction conditions in the reactor are shown in Table XVI (*18*).

TABLE XVI

REACTOR CONDITIONS FOR OXIDIZING ACETALDEHYDE TO ACETIC ANHYDRIDE

Temperature	56°C
Catalyst conc.	0.013–0.015 moles liter^{-1}
Acetic anhydride	1.4 moles liter^{-1}
Acetic acid	6.1 moles liter^{-1}
Water	1.24 moles liter^{-1}
Isopropyl acetate	3.4 moles liter^{-1}
Acetaldehyde	2.0 moles liter^{-1}

The acetic anhydride production rate was 2–5 gm moles per liter per hour.

In a continuous process, the pressure is 1.2 to 2 atm absolute and air is mixed with recycled gas so the oxygen content going to the reactor is 7–9% and the rate is 3.5–5 m^3 per liter of reaction mixture per hour. The heat of reaction is removed by cooling coils in the reactor. The exit gas is cooled to 10°C to remove reaction products and a part recycled. Volatile acetaldehyde in the recycle gas goes back to the reactor. That left in the gas to be vented is recovered. Liquid product is removed from the reactor continuously to recover the products, and acetaldehyde is fed continuously. 95% of the acetaldehyde is converted to acetic anhydride plus acetic acid, and the anhydride content is 50–80%. Oxidation byproducts, e.g., ethylidene diacetate, can be used as part of the solvent to protect the acetic anhydride from the water of reaction (*22*).

X. Acetic Acid from Methanol and Carbon Monoxide

Acetic acid can be made by the following type of reaction:

$$CH_3OH + CO = CH_3COOH$$

Dimethyl ether, a byproduct from methanol synthesis, can replace some of the methanol. The precursor of the catalyst is cobalt iodide which, under reaction conditions, is converted to cobalt hydrocarbonyl and hydrogen iodide. The reaction conditions are 250°C and 650 atm pressure. As would be expected under these conditions, methanol reacts with acetic acid to produce methyl acetate. So water is added to the reaction mixture to keep the concentration of methyl acetate under control. The concentration of the several components in the reactor approximates 30% methyl acetate, 30% acetic acid, 30% water, and 10% methanol. Methyl acetate, catalyst, and minor byproducts are recycled to the reactor. The yield of acetic acid is 90% based on the methanol.

As might be expected, the hot reaction mixture is highly corrosive so special metallurgy must be used for the reactor (*7*, *24*).

XI. Olefins to Neoacids

This reaction may be represented:

$$C_nH_{2n} + CO + H_2O = C_nH_{2n+1}COOH$$

in which the acid formed has the "neo" structure, i.e., contains a quaternary carbon atom:

$$R_3CCOOH$$

The olefin, e.g., *i*-C_4H_8, reacts with CO and water at room temperature and 35–70 atm pressure of CO in the presence of acidic catalysts, e.g., H_2SO_4 of 80–90% strength or boron trifluoride dihydrate. With sulfuric acid, the acid to olefin ratio is greater than 1:1 and usually 2:1 or 3:1. The neoacid is isolated from the reaction mixture (*21*, *43*).

Neopentanoic (pivalic acid), neoheptanoic, neodecanoic and neotridecanoic acids are being produced commercially (*43*) using boron trifluoride dihydrate as the catalyst (*21*).

XII. Glycol to Glyoxal

In this process glycol is oxidized to glyoxal with air at 345°C and 10 atm pressure. A small amount of a halide (presumably ethylene dichloride) is added to partly poison the catalyst surface and inhibit further oxidation. The catalyst is 3–8% CuO via copper nitrate deposited on inert, refractory alumina (*14*).

A. Catalyst Supplier

Harshaw Chemical Co.

Cu–1900 S: 5% CuO on high fired inert alumina, ½ inch spheres. The ABD is 1.2, the surface area 1 m^2gm^{-1}, and the pore volume 0.07 cm^3gm^{-1}. This catalyst was not designed for this service but the researcher may find it a convenient place to start if the particle size is not too large.

XIII. Furfural to Furoic Acid

This process may be obsolete but it employs some techniques that may be of use. The catalyst is silver but employed in an unusual way.

A calcium hydroxide slurry is prepared containing enough alkalinity to neutralize all the furoic acid formed. Silver nitrate is added to deposit the catalyst *in situ*. The hot mixture with furfural added is oxidized with air until most of the furfural is consumed. Unreacted furfural is stripped with steam and the solution of calcium furoate filtered from any residue. The furoic acid is recovered by neutralization (*26*).

XIV. Mercaptans to Disulfides ("Sweetening")

Mercaptans in gasoline have an objectionable odor ("sour" gasoline). By converting them to disulfides which have less odor, a "sweet" gasoline is produced; hence the term "sweetening." The process and name is applied to other petroleum products where mercaptans may cause problems other than odor.

There are many proprietary processes with proprietary catalysts. Only a few processes are described here to illustrate the principles. H_2S and methyl mercaptan, if present, are first removed by treating with a sodium hydroxide solution (*39*).

A. Inhibitor Sweetening

A *p*-phenylene diamine type antioxidant (with enough R groups on the nitrogens to make the product soluble in gasoline) is added to the gasoline to provide most if not all the ultimate antioxidant required for storage stability. Air is then added and the whole mixed thoroughly with 10–30% sodium hydroxide solution. After the caustic solution settles and is separated, the gasoline is stored until delivered. The sweetening reaction may still be proceeding slowly during storage and transit but the gasoline is sweet by the time it reaches the consumer (*3, 35*).

1. *Suppliers of Proven Catalysts*

E. I. du Pont de Nemours and Company, Inc., Ethyl Corporation, Tennessee Eastman Corporation, and Universal Oil Products Company all supply substituted *p*-phenylene diamines for this purpose.

B. Doctor Sweetening

This is one of the oldest petroleum refining processes. It may be represented by the equations:

$$2\,RSH + Na_2PbO_2 = PbR_2S_2 + 2\,NaOH$$

$$PbR_2S_2 + S = PbS + RSSR$$

The PbS is regenerated by air

$$PbS + 2\,O_2 + 4\,NaOH = Na_2PbO_2 + 2\,H_2O + Na_2SO_4$$

In essence, oxygen, sulfur, and sodium hydroxide are consumed, with lead compounds acting as catalysts to oxidize the mercaptans to disulfides (*4*).

C. Copper Sweetening

This is similar to doctor sweetening but $CuCl_2$ takes the place of sodium plumbite. The equations are:

$$2\,RSH + 2\,CuCl_2 = 2\,CuCl + RSSR + 2\,HCl$$

$$2\,CuCl + 2\,HCl + \tfrac{1}{2}\,O_2 = 2\,CuCl_2 + H_2O$$

The gasoline, naphtha, or kerosine from copper sweetening may contain traces of copper compounds which are powerful oxidation catalysts for these products. It is usual to wash the product with sodium sulfide to convert the Cu to harmless CuS. In addition, a few parts per million of metal deactivators to chelate the Cu may be added to the product. The copper chloride may be sorbed on a porous granular support and the sweetening reactions carried out in s fixed bed of the catalyst. Alternatively, the copper chloride may be sorbed on clay and the sweetening carried out as a slurry operation (*5*).

D. The Bender Process

The Bender process resembles a fixed bed version of the doctor process in that the reactions are the same. The "sour" jet fuel, kerosine, or home heating oil + air + a little sulfur is passed through fixed beds of catalyst. The catalyst beds are supported lead oxide–lead sulfide and are continuously regenerated by the air and sulfur (*6, 42*).

XV. Fuel Cells

As of this writing, reliable fuel cells have been developed for H_2–O_2 systems for space application. Catalysts are essential for both cathode and anode, and catalysts developed for H_2 are not suitable for other fuels. Because of the competitive nature of the developments, detailed catalyst information is not available.

Some cells employ an acidic ion-exchange membrane as the electrolyte with specially prepared platinum anodes and cathodes (*19*, *33*). Others employ KOH as the electrolyte. Besides Pt electrodes, many others have been studied, especially various silver combinations for the oxygen electrode and nickel combinations for the hydrogen electrode (*20*). While experimental cells have operated starting with hydrocarbons, ammonia and methanol fuels seem to be favored currently.

REFERENCES

1. Algy, J., Asselineau, L., Busson, C., Cha, B., and Sandler, H., *Hydrocarbon Process.* **47,** No. 12, 131 (1968).
2. Anonymous, *Chem. Eng. News* **39,** No. 16, p. 52 (1961).
3. Anonymous, *Hydrocarbon Process. Petrol. Refiner* **43,** No. 9, 209 (1964).
4. Anonymous, *Hydrocarbon Process. Petrol. Refiner* **43,** No. 9, 208 (1964).
5. Anonymous, *Hydrocarbon Process. Petrol. Refiner* **43,** No. 9, 203 (1964).
6. Anonymous, *Hydrocarbon Process. Petrol. Refiner* **43,** No. 9, 199 (1964).
7. Anonymous, *Hydrocarbon Process. Petrol. Refiner* **44,** No. 11, 160 (1965).
8. Anonymous, *Hydrocarbon Process.* **46,** No. 11, 135 (1967).
9. Anonymous, *Hydrocarbon Process.* **46,** No. 11, 137, 203 (1967).
10. Anonymous, *Hydrocarbon Process.* **46,** No. 11, 211 (1967).
11. Arganbright, R. P., and Evans, R. J., *Hydrocarbon Process. Petrol. Refiner* **43,** No. 11, 159 (1964).
12. Bashkirov, A. N., and Kamzolkin, V. V., *Proc. 5th World Petrol. Congr., New York* **4,** No. 15, 175 (1959).
13. Bashkirov, A. N., Sokolova, S. M., Kamzolkin, V. V., and Andreyeva, T. P., *Proc. 6th World Petrol. Congr., Frankfurt, Main* **4,** No. 38 (1963).
14. Bohmfalk, J. M., McNamee, R. W., and Barry, R. P., "Modern Chemical Processes," Vol. 2, p. 55. Reinhold, New York, 1952.
15. Brill, W. F., *Ind. Eng. Chem.* **52,** 837 (1960); U.S. Patent 2,853,514 (September 23, 1958).
16. Broich, F., and Groseman, H., *Erdöl Kohle* **18,** 360 (1965).
17. Burney, D. E., Weisemann, G. H., and Fragen, N., *Petrol. Refiner* **38,** No. 6, 186 (1959).
18. Carpenter, B. H., *Ind. Eng. Chem. Process Design Develop.* **4,** 105 (1965).
19. Cohn, J. G., *Englehard Ind. Tech. Bull.* **9,** No. 1, 5 (1968).
20. Dittmann, H. M., Justi, E. W., and Winsel, A. W., *Am. Chem. Soc. Div. Petrol. Chem. Preprints* **6,** No. 4, B-139 (1961); *Fuel Cells* **2,** 133 (1961).
21. Ellis, W. J., and Roming, C., Jr., *Hydrocarbon Process. Petrol. Refiner* **44,** No. 6, 139 (1965).

22. Gobron, G. L., and Alheritiere, L. A., *Chem. Eng. Progr.* **60,** No. 9, 55 (1964).
23. Helba, E. I., Dessau, R. M., and Koehl, W. J., Jr., *Am. Chem. Soc. Div. Petrol. Chem. Preprints* **14,** No. 2, A-44 (1969).
24. Hohenschutz, H., von Kutepow, N., and Himmele, W., *Hydrocarbon Process.* **45,** No. 11, 141 (1966).
25. Illingworth, G. E., and Lester, G. W., *Am. Chem. Soc. Div. Petrol. Chem. Preprints* **12,** No. 3, 161 (1967).
26. Isenhour, L. L., U.S. Patent 2,041,184 (May 19, 1936).
27. Kaeding, W. W., *Hydrocarbon Process. Petrol. Refiner* **43,** No. 11, 173 (1964).
28. Kaeding, W. W., Lindblom, R. O., Temple, R. G., and Mahon, H. I., *Ind. Eng. Chem. Process Design Develop.* **4,** 97 (1965).
29. Kochi, J. K., and Subramanian, R. V., *J. Am. Chem. Soc.* **87,** 4855 (1965).
30. Krekeler, H., and Krönig, W., *Proc. 7th World Petrol. Congr., Mexico City* **5,** 41 (1967).
31. Loder, D. J., U.S. Patent 2,245,528 (June 10, 1941).
32. Melchiore, J. J., Moyer, H. R., and Christman, L. J., *Advan. Chem. Ser.* **51,** 89 (1965).
33. Oster, E. A., and Chapman, L. E., *Am. Chem. Soc. Div. Petrol. Chem. Preprints* **6,** No. 4, B-79 (1961); *Fuel Cells* **2,** 168 (1961).
34. Robinson, R. E., U.S. Patent 3,190,912 (June 22, 1965).
35. Rosenwald, R. H., *Petrol. Process.* **11,** No. 10, 91 (1956).
36. Rouchaud, J., and Lutete, B., *Ind. Eng. Chem. Prod. Res. Develop.* **7,** 266 (1968).
37. Schwerdtel, W., *Hydrocarbon Process.* **47,** No. 11, 187 (1968).
38. Smidt, J., and Krekeler, H., *Proc. 6th World Petrol. Congr., Frankfurt, Main* **4,** 461 (1963); *Hydrocarbon Process. Petrol. Refiner* **42,** No. 7, 149 (1963).
39. Tait, T., *Advan. Chem. Ser.* **5,** 151 (1951).
40. Technical Bulletin "Witco Metallic Carboxylates." Organics Division, Witco Chemical Corp., New York, 1966. (Copies are available from the supplier.)
41. Towle, P. H., and Baldwin, R. H., *Hydrocarbon Process. Petrol. Refiner* **43,** No. 11, 149 (1964).
42. Waterman, L. C., and Wiley, R. A., *Petrol. Refiner* **34,** No. 9, 182 (1955).
43. Wickson, E. J., and Moore, R. R., *Hydrocarbon Process. Petrol. Refiner* **43,** No. 11, 185 (1964).

CHAPTER 20

OXIDATION (ORGANIC) VAPOR PHASE

I. Naphthalene or o-Xylene to Phthalic Anhydride—Benzene to Maleic Anhydride

The earliest catalytic process for making phthalic anhydride involved heating naphthalene in concentrated sulfuric acid using Hg as the catalyst. Sulfuric acid was the oxidizing agent. The phthalic anhydride sublimed out of the reaction mixture. While this is no longer a commercial process, it is a catalytic technique that might be useful.

Currently, phthalic anhydride is made by oxidizing naphthalene in the presence of a supported vanadium pentoxide catalyst. *o*-Xylene can be used in place of naphthalene, although most plants are designed for one or the other, there being a few plants that can use either. Maleic anhydride is made in the same way from benzene, so these processes will be treated together. While the principles are the same, operating conditions for recovery of the product and optimum yields vary.

In oxidizing naphthalene rather than *o*-xylene, there are two carbon atoms to be oxidized to CO_2 with naphthalene; with *o*-xylene, only hydrogens must be oxidized. With less heat liberated in the reactor, *o*-xylene should give better yields. In practice there has been a running battle between *o*-xylene and naphthalene with improvements in catalysts and operating techniques for both. Both are being used and competing. Details of many of the catalyst improvements are proprietary so the catalysts listed below may not incorporate all the latest advances (*23*).

These oxidations are highly exothermic so heat removal is essential.

Several techniques have been adopted. One uses catalyst tubes with a molten salt heat transfer fluid outside the tubes. Another uses a fluidized catalyst with cooling coils in the catalyst bed.

In operation, air at 0.5 atm pressure is heated to 135°–150°C and saturated with naphthalene or *o*-xylene so that the air:hydrocarbon mixture is below the explosion limit, e.g., air:hydrocarbon 20–30:1. The mixture is then passed through catalyst pellets inside the tubes of a heat exchanger with a cooling medium outside the tubes. The temperature is 315°–375°C or even higher. In some respects the reaction resembles burning coke off a catalyst; i.e., the major heat release tends to be in a limited part of the catalyst bed. Recycle of part of the exit gas is sometimes used to spread out the oxidation zone and obtain better heat control (*14*).

The fluidized catalyst bed by its movement intrinsically tends to spread out the burning zone. Once the reaction has started, it is not necessary to preheat the air or hydrocarbon feed. By bringing these in cold, some of the heat of reaction is absorbed. Also, since they can be fed separately, a greater flexibility is available in the air:hydrocarbon ratio, i.e., well into the explosion region is possible. A common range for the fluidized catalyst reactor is 10–15:1 air:hydrocarbon ratio. The retention time of the reactants in the reactor is 10–15 seconds (*9*, *35*).

The hydrocarbon is completely consumed with some of it being oxidized to CO_2 and H_2O. There is some byproduct formation; from naphthalene, some quinone plus maleic and benzoic acids. *o*-Xylene gives more maleic and benzoic acids with some phthalide (*35*). Phthalic anhydride yields are 93–97 weight percent of the original hydrocarbon (*14*), with some of the recent plants doing better.

In some respects, the catalytic part of a phthalic anhydride plant is a minor part. Phthalic anhydride sublimes readily so the exit gases from the reactors tend to carry significant quantities of phthalic anhydride along with them. The settling chambers, etc., for recovering this phthalic anhydride can be large, expensive to operate, and clean. Exit gases must be scrubbed or burned, for traces of phthalic anhydride or maleic anhydride are objectionable air pollutants.

While the catalysts are based on vanadium pentoxide, some molybdenum oxide is added to the catalyst for making maleic anhydride.

Phthalic anhydride is readily made by a vapor phase process because it is volatile and readily leaves the catalyst bed. Isophthalic or terephthalic acids are far less volatile and hence stay in the catalyst bed and are oxidized further. For this reason liquid phase oxidation processes (Chapter 19) are used for them. Liquid phase processes can be used for phthalic also and some is made from mixed xylenes. However, by far most of the phthalic anhydride is made by the vapor phase process.

In making maleic anhydride the yield is about 75–80 weight percent plus 5–10% fumaric acid based on the benzene (*6, 16*).

A. Suppliers of Proven Catalysts

American Cyanamid Company

Aero Catalyst PAA-3: 62.0% SiO_2, 6.1% V_2O_5, 10.8% K_2O, and 17.0% SO_3, with a mole ratio of $SO_3:K_2O$ of 1.85, as a powder 95% finer than 149 μ. The ABD is 0.55 and the surface area 220 m^2gm^{-1}.

Davison Chemical Division of W. R. Grace & Co.

Grade 902: 9% V_2O_5, 29% K_2SO_4, 12% SO_3, SiO_2 50%, as powder. The ABD is 0.6, the surface area 40 m^2gm^{-1}, and the average pore diameter 30 Å.

Grade 906: 4% V_2O_5, 14.9% K_2SO_4, 6.7% SO_3, SiO_2 73%, as powder. The ABD is 0.6, the surface area 200 m^2gm^{-1}, and the average pore diameter 95 Å.

Harshaw Chemical Company

V-0301 T: 10% V_2O_5 supported on activated alumina, as ⅛ inch tablets. The ABD is 1, the surface area 78 m^2gm^{-1}, and the pore volume 0.33 cm^3gm^{-1}.

V-0501 S: 10% V_2O_5 supported on inert alumina, as ¼ inch spheres. The ABD is 1.3 and the surface area 1 m^2gm^{-1}.

V-0601 T: 10% V_2O_5 mounted on high activity alumina, as ⅛ inch tablets. The ABD is 1, the surface area 115 m^2gm^{-1}, and pore volume 0.28 cm^3gm^{-1}.

V-0701 T: 10% V_2O_5 mounted on a silica-containing alumina, as ⅛ inch tablets. The ABD is 0.9, the surface area 139 m^2gm^{-1}, and the pore 0.45 cm^3gm^{-1}.

V-1001 E: 5% V_2O_5 + 5% MoO_3 on inert alumina, as 3/16 inch extrudates. The ABD is 1.3 and the surface area 1 m^2gm^{-1}.

V-1002 E: 6% V_2O_5 + 3% MoO_3 on inert α-alumina, as 3/16 inch extrudates. The ABD is 1.3 and the surface area 1 m^2gm^{-1}.

II. Ethylene to Ethylene Oxide

The catalytic oxidation of ethylene to ethylene oxide has virtually displaced the older chlorohydrin process. A large part of this success is due to the development of more and more selective catalysts so that less of the ethylene is oxidized to the carbon oxides and water. As a result, the developers of these catalysts consider them proprietary and part of what the licensee pays for.

In general, the reaction is carried out in the temperature range of 200°–300°C and at 10–30 atm pressure in a fixed catalyst bed, where the basis of the catalyst is silver oxide mounted on a refractory inactive support. Ethylene and air are mixed with recycle gas containing mainly nitrogen and unconverted ethylene and the mixture passed over the catalyst. The catalyst is contained in tubes in a heat exchanger with a boiling cooling fluid on the outside of the tubes to absorb the heat of reaction and control the temperature. The ethylene oxide is scrubbed from the gas and part of the gas is recycled. The rest is passed to a second stage reactor to convert the residual ethylene. After scrubbing to recover ethylene oxide, the nitrogen and CO_2 are discarded (*8*, *10*).

In an alternate process, oxygen is used instead of air. The oxygen–ethylene mixture is always on the rich side of the explosion limit, e.g., $<5\%$ O_2. Since the recycle stream is mainly C_2H_4, a second stage to consume residual C_2H_4 in the nitrogen stream is not necessary; there is no nitrogen (*11*).

Yields are in excess of 100% by weight based on the ethylene. The selectivity (less C_2H_4 oxidized to CO_2) is improved by adding a few parts per million of an organic chloride, e.g., ethylene dichloride, to the feed (*26*, *31*). Apparently, Ag chemisorbs the Cl^-, partly poisoning the Ag for O_2 adsorption and improving selectivity in this way (*33*).

For laboratory or pilot plant exploration, a temperature in the range of 250°–280°C is suitable with an air to ethylene ratio of 10:1 and a contact time of 1–5 seconds. A catalyst useful for exploratory work may be prepared as follows: 70 gm of 8 mesh corundum, 22 gm of silver oxide and 2.25 gm of barium peroxide are mixed with 100 ml of water; stir the mixture well while heating it to evaporate the water. The dried product is then heated in air at 115°C for 10 hours before use (*31*).

A. Catalyst Suppliers

None of the regular catalyst suppliers offer this type of catalyst at the time of this writing.

III. Higher Olefin Oxides

Attempts to use the silver catalysts with higher olefins to make the oxides by direct vapor phase oxidation have so far resulted in too much complete oxidation to CO_2. The chlorohydrin process has persisted for propylene oxide for example and many of the chlorohydrin plants originally designed for ethylene oxide were adapted to make propylene oxide.

Catalysts have been reported for the known hydroperoxide reaction with olefins (*30*, *36*).

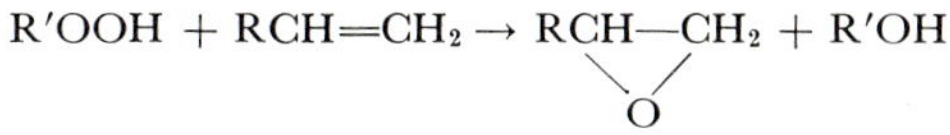

The catalysts are compounds of molybdenum or tungsten with several other elements being active (*30*). An inexpensive hydrocarbon is partly converted to the hydroperoxide by air, the olefin hydrocarbon dissolved in this mixture, and the catalyst in hydrocarbon soluble form, e.g., naphthenate, added. The mixture is then heated to effect reaction. It is thought that the reaction is ionic in nature rather than free radical, the type usually associated with hydroperoxides (*30*).

Olefins vary in their reactivity in this type of reaction with ethylene reacting slowest and propylene next. Higher olefins react still faster. The following range of conditions were given for propylene. Temperature 80°–130°C, reaction time 0.3–3 hours, pressure 17–70 atm, olefin concentration 2–6 moles per mole of hydroperoxide and 0.001–0.006 moles of catalyst per mole of hydroperoxide (*30*).

Olefin oxides can also be formed by the direct reaction of the olefin with oxygen in the presence of a hydrocarbon soluble cobalt catalyst (*4*). The reaction takes place at atm pressure and 50°–100°C. In contrast with the preceeding reaction, this one apparently takes place by a free radical mechanism.

Olefin oxides can also be formed with some olefins and oxygen in the absence of added catalysts (120°C and 20 atm pressure) (*22*).

So far as the author knows, neither of the two previous reactions are being utilized commercially at this time to make olefin oxides.

A. Suppliers of Proven Catalysts

Climax Molybdenum Company supplies molybdenum naphthenate and molybdenyl bis-acetylacetonate.

The following companies supply hydrocarbon soluble forms of cobalt.

Carlisle Chemical Works, Inc.
Harshaw Chemical Company
MacKenzie Chemical Works, Inc.
Nuodex Division of Tenneco Chemicals, Inc.
Organics Division, Witco Chemical Company

IV. Alcohols to Aldehydes or Ketones

A. Methanol to Formaldehyde

There are two rather different processes for oxidizing methanol to formaldehyde. In one, silver is the catalyst and the process is really an oxidative dehydrogenation. In the other, an iron promoted molybdena catalyst is used and the process is oxidation (*40*).

In the silver process, the catalyst may be a silver metal gauze or may be particles of silver plus a support. The process operates well on the rich side of the explosion limit, e.g., 30–40% CH_3OH vapor in air. Steam may be introduced by vaporizing a mixture of water and methanol, or it may be introduced separately. Oxidation begins slowly at 300°–350°C and rapidly accelerates due to the rising temperature from the exothermic reaction in the adiabatic catalytic reactor. As the temperature rises, endothermic dehydrogenation takes place helping to control the temperature rise. The maximum temperature is 600°–650°C and the air:methanol ratio is controlled to maintain this temperature. The formaldehyde is unstable at the maximum reaction temperature of 600°–650°C so the reaction products are quenched rapidly. Otherwise, the formaldehyde decomposes into $CO + H_2$.

Since the recovery of methanol from the product for recycle is expensive, efforts are made to obtain high conversion in a single pass through the catalyst. Conversions are upwards of 90% per pass, with yields of 80–90+% formaldehyde based on the methanol converted (*12*, *27*). The formaldehyde is produced as a 40% solution containing about 1.7% methanol, i.e., the unreacted methanol, about 0.01% formic acid, and the rest is water.

The exit gas contains 17–20% H_2 (from the dehydrogenation), nitrogen, and a little formaldehyde. It is usually burned so the formaldehyde does not pollute the air.

Copper catalysts have been used in the past but these have largely disappeared for this use.

With the iron–molybdena catalyst, the air:methanol mixture is on the lean side of the explosion limit and the reaction is entirely one of oxidation. The following laboratory conditions are given for a catalyst containing equiatomic proportions of Fe and Mo as oxides (*3*). Temperatures of 350°–400°C (375°C optimum) and a feed of 95 liters of air + 9 gm hour^{-1} of methanol for 45 ml of catalyst are used. The average yield of formaldehyde was 95% with the rest going to CO.

An alternate catalyst contains 17.5% Fe_2O_3 + 82.5% MoO_3 (weight) (*29*).

Commercially, the catalyst is put inside the tubes of a heat exchanger with a cooling fluid outside to control the heat of reaction. Again, the objective is to obtain essentially complete conversion without undue decomposition of the formaldehyde. Since no water or steam is ordinarily added in the oxidation, the concentration of formaldehyde in the product can be controlled by the amount of water added in the absorber. Commercial solutions up to 60% formaldehyde are available (*13*).

Catalysts for maximum activity contain more molybdenium than the 1:1 Fe:Mo mentioned above, i.e., even somewhat more than that required for forming ferric molybdate, $Fe_2(MoO_4)_3$ (*20*, *29*). There is evidence that the main constituent of the catalyst as prepared is a ferric molybdate (*20*).

1. *Catalyst Poisons*

The silver is poisoned by volatile chlorine and sulfur compounds. Volatile iron compounds or finely divided iron rust particles deposited on the catalyst damage it.

2. *Supplier of Proven Catalyst*

Harshaw Chemical Company

Ag–0101 E: 3.5 to 4% Ag on an inert alumina support, as ⅛ inch extrudates. The ABD is 1.3 and the surface area 1 m^2gm^{-1}.

The author does not know of a supplier for the iron oxide–molybdenum oxide catalysts.

B. Higher Alcohols

Ethanol and air passed over the silver catalysts at 550°–570°C give 50–55% conversion per pass to acetaldehyde with an 85% yield. The process is operated in the rich zone to avoid explosions. Since there are more economical ways of making acetaldehyde, this one is rarely used commercially except for small quantities where the convenience outweighs the higher cost of ethanol.

Isopropanol gives acetone and other higher alcohols react over the silver catalysts but catalytic dehydrogenation (Section VI, Chapter 6) is ordinarily used instead of oxidation.

V. Dimethyl Ether to Formaldehyde

In making synthetic methanol, about 2% dimethyl ether is produced and, as of this writing, has little market. There is a process for oxidizing it to formaldehyde. The catalyst is WO_3 + 10% phosphoric acid on an inert carrier, e.g., silicon carbide or α-alumina. The dimethyl ether:air mixture is leaner than the explosion limit (3.4%). The temperature is 450°–530°C and the GHSV 1000–4000 $V_gV_c^{-1}$ hr^{-1} at atmospheric pressure. Conversion is 90–100% per pass with yields of 70–80% of theory (*37*).

VI. *n*-Butenes to Butadiene

The dehydrogenation of hydrocarbons is a strongly endothermic reaction. In addition, the hydrogen formed limits the concentration of dehydrogenated product that can be obtained before equilibrium is reached. These two effects have inspired chemists and chemical engineers to seek an oxidation process that would solve both problems.

Both 1-C_4H_8 and 2-C_4H_8 may be oxidized over a "bismuth molybdate" catalyst to give butadiene, although the reaction with 2-butene is considerably slower. Temperatures are in the 400°–500°C range. For example, 80% conversion of 1-butene was obtained at 460°C, GHSV 3600 $V_g V_c^{-1}$ hr^{-1}, and atmospheric pressure (*2*). The selectivity is good; 90+% in the 60+% conversion range.

The so called "bismuth molybdate" has been the subject of considerable study seeking to determine the active species. Two observations stand out: (*a*) the maximum activity occurs at a Bi:Mo ratio between 1:1 and 2:1 (*19*); (*b*) specific activity is not necessarily related to crystallinity (*21*).

Apparently, the butene reacts with a catalyst oxygen to yield butadiene. The catalyst is reoxidized by the oxygen supplied in the reaction mixture (*18*).

Oxygen can be added to the catalysts used for the steam:butene dehydrogenation (Section II–A, Chapter 6) reacting under normal operating conditions or at slightly reduced temperatures, and some oxidative dehydrogenation effected.

VII. Propylene or Isobutylene to Unsaturated Aldehydes

Propylene treated under the butene to butadiene oxidation conditions gives acrolein. Similarly, isobutylene gives methacrolein (*1*, *2*, *38*) using the "bismuth molybdate" catalyst.

Cuprous oxide is also a catalyst for the oxidation of propylene to acrolein. Alkyl bromides in the feed, e.g., 0.05% methyl bromide, improves the selectivity and reduces complete oxidation (*24*, *41*).

VIII. Higher Olefins to Dienes

All three of the methylbutenes give isoprene with selectivities of 70–80%, rather than the upwards of 90% observed with the butenes, when subjected to oxidative dehydrogenation with the "bismuth molybdate" catalyst. More oxygen containing products are obtained than with butene (*1*).

The aliphatic olefins C_5–C_7 give dienes, but the ratio of diene to oxygenated product varies widely with the structure of the olefin (*1*). Interestingly, cyclopentene gave only CO plus CO_2.

IX. Suppliers of Proven Catalysts

The author is not aware of a commercial supplier of the "bismuth molybdate" catalyst at this time. The following laboratory preparation may be useful (*19*). Freshly prepared and washed $BiO(OH) \cdot H_2O$ and H_2MoO_4 (undried) are mixed in the desired ratio, dispersed in water, and boiled 20 hours. The product is filtered, dried at 110°C overnight, and calcined at 500°C for 2 hours.

Harshaw Chemical Company

Cu–0307 T: 99% CuO as ⅛ inch tablets. The ABD is 1.3, the surface area 5 m^2gm^{-1}, and the pore volume 0.26 cm^3gm^{-1}. This must be reduced to the cuprous form by heating in hydrogen at 300°–350°C until the catalyst begins to turn red.

X. Ammonia Plus Methane to Hydrogen Cyanide

This process is known as the Andrussow process. L. Andrussow did brilliant work with Max Bodenstein unravelling the complex kinetics of ammonia oxidation on Pt gauze. With this background, he conceived the idea of adding methane to the system to produce HCN. The reaction can be represented:

$$NH_3 + CH_4 + 1.5\ O_2 = HCN + 3\ H_2O$$

The catalyst is Pt with 1–10% Rh or Pt with 2–3% Ir, and is in the form of gauze, just as in the ammonia oxidation process. The temperature is 1000 ± 100°C at about 0.5 atm pressure. The feed is 10–12% NH_3, 13% CH_4 (CH_4:NH_3 = 1.05), 75% air at 200 to 400 liters per hour per square centimeter of gauze. The linear velocity is 0.6–1.2 m sec^{-1}. The reaction is strongly exothermic and is self-sustaining (*5, 25*). The HCN yield is 60–65% with 10–20% of the NH_3 lost as N_2 and the rest unconverted. The unit is operated for complete conversion per pass and no attempt is made to recover NH_3 and recycle it. HCN has a strong tendency to polymerize, so the reaction mixture must be quenched quickly and appropriate inhibitors added in the recovery system to prevent this.

An example of the composition of the reactor gas is given in Table XVII.

TABLE XVII

OXIDATION OF NH_3 PLUS CH_4 TO PRODUCE HCN; REACTOR GAS COMPOSITION

Component	Mole %
HCN	5.9
NH_3	1.6
CO_2	0.2
CO	4.4
CH_4	0.5
H_2	7.5
O_2	0.1
N_2	56.7
H_2O vapor	23.1

As alternatives to Pt gauze, it is also possible to use Pt on refractory supports as the catalyst.

A. SUPPLIERS OF PROVEN CATALYSTS

Engelhard Industries, Inc. or Matthey Bishop, Inc. supply Pt gauze catalysts, as well as Pt on refractory supports.

XI. Ammonia Plus Propylene to Acrylonitrile

Propylene + ammonia + air is reacted in a fluidized catalyst reactor at 0.5 to 2 atm pressure and 400°–500°C to produce primarily acrylonitrile. Two catalysts are used; one is the "bismuth molybdate" type described above. The other is a silica supported "uranyl antimonate" type (*15*, *39*). The yield is around 0.7–0.75 kg of acrylonitrile per kilogram of propylene. The byproducts are 0.15–0.20 kg of HCN and 0.10 kg of acetonitrile per kilogram of acrylonitrile, with the rest going to CO and CO_2. The "uranyl antimonate" catalyst gives more acrylonitrile and less acetonitrile than the "bismuth molybdate." With both catalysts, the unit is operated to give essentially complete conversion per pass.

XII. Isobutylene Plus Ammonia to Methacrylonitrile

Methacrylonitrile can be produced from isobutylene, ammonia, and air, using the same catalysts and conditions given above. A cuprous oxide catalyst has also been used (*32*).

XIII. Catalyst Suppliers

The catalysts are available to lisencees of the process. For exploratory work in the laboratory, the author suggests the "bismuth molybdate" described above in Section IX or the preparation published for oxidizing propylene to acrolein (*38*). Detailed laboratory preparations are given for both the silica supported and the unsupported "uranyl antimonate" catalyst (*28*).

XIV. HCl to Chlorine

The Deacon process in which HCl was catalytically oxidized to chlorine and water suffered because an equilibrium was involved and the product was a mixture of chlorine, water, oxygen, and HCl. By including an organic substance in the mixture so that it reacts with the chlorine as formed, it is possible to displace the equilibrium and to chlorinate organic molecules with HCl as the source of the chlorine. Several organic processes are based on this broad principle.

XV. Benzene to Chlorobenzene

In the Raschig process, benzene, HCl, water vapor (from aqueous HCl), and air are passed over mixed Cu and Fe chlorides on a porous support. The pressure is near atmospheric and the temperature is 210°–230°C. The conversion of benzene to chlorobenzene is 10% per pass. As might be expected, the reaction is exothermic and both feed and products are corrosive (*7*, *34*). The author does not know a catalyst supplier, but if he wanted to experiment in this area he would start with 5% $CuCl_2$ + 5% $FeCl_3$ mounted on kieselguhr.

In the second stage of the Raschig process, chlorobenzene is hydrolyzed to phenol + aqueous HCl in the presence of a silica gel catalyst at about 500°C. The aqueous HCl is recycled to the chlorination step.

XVI. Oxychlorination or Oxyhydrochlorination

In one process for making vinyl chloride, C_2H_4 is reacted with Cl_2 to make ethylene dichloride. This is pyrolyzed to vinyl chloride and HCl. Unless C_2H_2 is available to react with the HCl to form more vinyl chloride, the HCl is lost to the vinyl chloride process. However, oxychlorination is finding a growing application for utilizing this HCl in a modified Deacon process to make more ethylene dichloride. The reaction can be represented

$$C_2H_4 + 2\,HCl + \tfrac{1}{2}\,O_2 = C_2H_4Cl_2 + H_2O$$

Since the Cl_2 is consumed as formed, the equilibrium can be driven to substantial completion and all HCl consumed in each pass through the catalyst. The catalysts are of the Deacon $CuCl_2$ type with proprietary modifications.

The operation is conducted in a fluidized solids bed for temperature control. Either O_2 or air can be (and both are) used. Ordinary $CuCl_2$ on a refractory nonadsorptive support becomes sticky at reaction temperature as the $CuCl_2$ approaches its melting point. As a result, more adsorptive supports (e.g., γ-alumina or silica gel) are used. Here the support must be hard enough so it does not attrite too rapidly. Otherwise, catalyst fines are lost and must be replaced.

Operating temperature is in the range of 325°–425°C and pressure is 2–10 atm. Utilizing all the HCl implies that C_2H_4 and O_2 are in excess. In any event, a small proportion of the C_2H_4 is oxidized (*17*).

Saturated hydrocarbons can be chlorinated in an analogous way.

A. Catalyst Suppliers

Most of the catalysts are proprietary and available to licensees only.

Harshaw Chemical Company

Cu–0905 T: 10% $CuCl_2$ mounted on activated alumina, as ⅛ inch tablets. The ABD is 1, the surface area 59 m^2gm^{-1}, and the pore volume 0.26 cm^3gm^{-1}.

Cu–0905 P: 10% $CuCl_2$ on activated alumina powder. The ABD is 1 and surface area 180 m^2gm^{-1}.

XVII. Oxidation to Carbon Dioxide

The ability of platinum to act as an oxidation catalyst for organic vapors, oxidizing them to CO_2, has been known for many years. Uses for this catalytic property were described in Chapter 12.

This catalytic property can be used in more direct ways and the following application merely illustrates this use. A small stream of natural gas is bled from a natural gas pipeline buried in the ground. This is brought to the catalytic unit mounted above ground on a pole. The natural gas is mixed with air and burned catalytically (flameless combustion) on the Pt surface. A solid-state thermoelectric generator uses the heat to generate an electric current which is applied to the metal pipe line to provide cathodic protection against corrosion. The units operate unattended in isolated locations.

A. Suppliers of Proven Catalysts

Both Engelhard Industries and Matthey Bishop, Inc. supply Pt catalysts.

REFERENCES

1. Adams, C. R., *Proc. 3rd Intern. Congr. Catalysis, Amsterdam, 1964* **1,** 240 (1965).
2. Adams, C. R., Voge, H. H., Morgan, C. Z., and Armstrong, W. E., *J. Catalysis* **3,** 379 (1964).
3. Adkins, H., and Peterson, W. R., *J. Am. Chem. Soc.* **53,** 1512 (1931).
4. Amir, E. M., *Am. Chem. Soc. Div. Petrol. Chem. Preprints* **8,** No. 1, 33 (1963).
5. Anonymous, *Hydrocarbon Process. Petrol. Refiner* **42,** No. 11, 182 (1963).
6. Anonymous, *Hydrocarbon Process. Petrol. Refiner* **44,** No. 11, 237 (1965).
7. Anonymous, *Hydrocarbon Process. Petrol. Refiner* **44,** No. 11, 256 (1965).
8. Anonymous, *Hydrocarbon Process. Petrol. Refiner* **44,** No. 11, 212 (1965); *Hydrocarbon Process.* **46,** No. 11, 175 (1967).
9. Anonymous, *Hydrocarbon Process.* **46,** No. 11, 214 (1967).
10. Anonymous, *Hydrocarbon Process.* **46,** No. 11, 176 (1967).
11. Anonymous, *Hydrocarbon Process.* **46,** No. 11, 177 (1967).
12. Anonymous, *Hydrocarbon Process.* **46,** No. 11, 181 (1967).
13. Anonymous, *Hydrocarbon Process.* **46,** No. 11, 182 (1967).
14. Anonymous, *Hydrocarbon Process.* **46,** No. 11, 213 (1967).
15. Anonymous, *Hydrocarbon Process.* **46,** No. 11, 141 (1967).
16. Anonymous, *Hydrocarbon Process.* **46,** No. 11, 199 (1967).
17. Anonymous, *Hydrocarbon Process.* **46,** No. 11, 239–243 (1967).
18. Batist, P. A., Kapteijns, C. J., Lippens, B. C., and Schuit, G. C. A., *J. Catalysis* **7,** 33 (1967).
19. Batist, P. A., Der Kinderen, A. H. W. M., Leeuwenburgh, Y., Metz., F. A. M. G., and Schuit, G. C. A., *J. Catalysis* **12,** 45 (1968).
20. Boreskov, G. K., *Proc. 3rd Intern. Congr. Catalysis, Amsterdam, 1964* **1,** 173 (1965).
21. Boutry, P., Montarnal, R., and Wrzyszcz, J., *J. Catalysis* **13,** 75 (1969).
22. Brill, W. F., *Advan. Chem. Ser.* **51,** 70 (1965).
23. Dixon, J. K., and Longfield, J. E., *Catalysis* **7,** 183 (1960).
24. Dixon, J. K., and Longfield, J. E., see reference 23, p. 250.
25. Dixon, J. K., and Longfield, J. E., see reference 23, p. 300.
26. Dixon, J. K., and Longfield, J. E., see reference 23, p. 237.
27. Gerloff, U., *Hydrocarbon Process.* **46,** No. 6, 169 (1967).
28. Grasselli, R. K., and Callahan, J. L., *J. Catalysis* **14,** 93 (1969).
29. Jiru, P., Wichterlova, B., and Tichy, J., *Proc. 3rd Intern. Congr. Catalysis, Amsterdam, 1964* **1,** 199 (1965).
30. Landau, R., Brown, D., Russell, J. L., and Kollar, J., *Proc. 7th World Petrol. Congr., Mexico City* **5,** 67 (1967); *Hydrocarbon Process.* **46,** No. 4, 141 (1967).
31. McBee, E. T., Hass, H. B., and Wiseman, T. A., *Ind. Eng. Chem.* **37,** 432 (1945).
32. Mann, R. S., and Rouleau, D., *Advan. Chem. Ser.* **51,** 40 (1965).

33. Meisenheimer, R. G., and Wilson, J. N., *J. Catalysis* **1,** 151 (1962).
34. Olive, T. R., *Chem. Met. Eng.* **47,** 770 (1940).
35. Spitz, P. H., *Hydrocarbon Process.* **47,** No. 11, 162 (1968).
36. Swern, D., *J. Am. Chem. Soc.* **69,** 1692 (1947).
37. Tadenuma, H., Murakami, T., and Mitsushima, H., *Hydrocarbon Process.* **45,** No. 11, 195 (1966).
38. Veatch, F., Callahan, J. L., Milberger, E. C., and Foreman, R. W., *Actes Congr. Intern. Catalyse, 2e, Paris, 1960* **2,** 2647 (1961).
39. Veatch, F., Callahan, J. L., Idol, J. D., Jr., and Newberger, E. C., *Chem. Eng. Progr.* **56,** 65 (1960).
40. Walker, J. F., "Formaldehyde." Reinhold, New York, 1953.
41. Wood, B. J., Wise, H., and Yolles, R. S., *J. Catalysis* **15,** 355 (1969).

CHAPTER 21

MISCELLANEOUS PROCESSES

I. Oxo Process (Hydroformylation)

The oxo process is the reaction of olefins with carbon monoxide plus hydrogen (1:1) to produce aldehydes. The aldehydes may then be hydrogenated to primary alcohols.

In its simplest form, propylene, hydrogen, and carbon monoxide are converted into a mixture of *n*- and isobutyraldehyde which are recovered and separated. Most of the plants convert "heptenes" (propylene–butylene copolymer) into "octyl alcohols," octenes (butylene dimers) into "nonyl alcohols," and propylene trimer into "decyl alcohols." Since the olefins are mixtures, so are the aldehydes and the alcohols.

The active catalyst is cobalt hydrocarbonyl, $HCo(CO)_4$, which is prepared *in situ*. It is usually formed from cobalt compounds soluble in the olefin, e.g., cobalt naphthenate, or from cobalt octacarbonyl.

The reaction is conducted in the liquid phase in which the cobalt hydrocarbonyl is soluble, i.e., a homogeneous catalyst. The operating temperature is 110°–190°C at 150–250 atm pressure. The cobalt concentration in the feed is 0.05–1% and typically 0.5%. The reaction is strongly exothermic. Cold $H_2 + CO$ may be used to control the temperature, e.g., gas may be recycled through the liquid in the reactor to effect cooling. The liquid reaction product (aldehydes + catalyst) is continuously withdrawn (*31*, *59*). Conversion of the olefin is typically 90+%, with only a small fraction of it hydrogenated to paraffin hydrocarbons.

Many methods for recovering the cobalt catalyst for recycle have been used. The cobalt hydrocarbonyl is thermally unstable without a high partial

partial pressure of CO, so the catalyst cannot be recovered by simple distillation (*33*). The following illustrates one method. The synthesis product containing the catalyst is extracted with an aqueous sodium carbonate solution when the water soluble salt of the catalyst (which is an acid) is formed:

$$2\,HCo(CO)_4 + Na_2CO_3 = 2\,NaCo(CO)_4 + CO_2 + H_2O$$

This recovers the catalyst from the product. The aqueous solution is treated with acid in the presence of the liquid olefin to be oxonated. The $HCo(CO)_4$ catalyst, which is only slightly soluble in water, is liberated and is extracted by the hydrocarbon (*33*) ready for reuse.

The aldehydes formed are hydrogenated in a fixed bed using a cobalt or nickel catalyst at 100°–150°C and 10–50 atm pressure.

There are several interesting features to the oxo process. The incoming aldehyde group ("formaldehyde," hence the term "hydroformylation") may add to either side of an unsymmetrical olefin:

$$RCH{=}CH_2 + CO + H_2 \nearrow RCH_2{-}CH_2{-}CHO \qquad (7)$$

$$RCH{=}CH_2 + CO + H_2 \searrow RCH(CHO){-}CH_3 \qquad (8)$$

The net result is that each olefin may produce two aldehydes and hence two primary alcohols. Since the olefin feed is usually a complex mixture of olefins, the alcohols are usually complex.

The ratio of Eq. (7) to Eq. (8) is subject to some control. Equation (7) tends to be favored by low operating temperatures while Eq. (8) is favored by high (190°–200°C) reaction temperatures. Equation (8) is favored by increased catalyst concentration. Typically the product contains 60–70% of products from Eq. (7).

Equation (8) is difficult with olefins in which one of the carbon atoms of the double bond is a tertiary carbon atom. Such a reaction forms a quaternary carbon atom. Accentuating the reaction conditions favoring Eq. (8) in studies with isobutylene gave ratios of 3-methylbutyraldehyde to pivaldehyde of 15–20:1. Temperatures to 220°C and pressures of 400 atm were used (*30*).

A. Catalyst Improvements

As this is being written, quite significant changes are taking place in the oxo process and catalysts. Replacing one CO ligand in the catalyst with a tertiary organophosphine, e.g., tributylphosphine or tricyclohexylphosphine, changes the reaction and illustrates the potent effect of the ligands in homogeneous catalysts. The new catalysts are more selective for producing the normal [Eq. (7)] product so that the ratio of A:B is 8–9:1 (*16, 23, 28, 48, 56*). Under mild conditions, it was shown that the reaction rate with $HCo(CO)_4$ was quite a bit faster than that with the phosphine ligand catalyst. When conditions were adjusted to get high olefin conversion, alcohols were the predominant products rather than the aldehydes (*56*). The catalyst also hydrogenated the aldehydes.

B. New Catalysts

Substituting Rh for Co in the catalyst system produces further interesting changes. One such catalyst is $[(C_6H_5)_3P]_2Rh(CO)Cl$ (*18, 20, 49*). The catalyst is reported to be more active in that concentrations of 10^{-3} to 10^{-4} M are adequate permitting reaction at 100°–125°C. Lower pressures can be used, 35 atm, and good selectivity obtained, i.e., 75–80% of the preferred Eq. (7) product.

In contrast to the thermally unstable $HCo(CO)_4$, the Rh complex is stable, permitting easier catalyst recovery. The reaction products can be distilled from the catalyst, recovering the catalyst directly for recycle (*18*). Reportedly, triphenylphosphine aids in the selectivity so that $5(C_6H_5)_3P$:Rh gives 100% selectivity. It also prevents oxidation of the catalyst since otherwise it is sensitive to oxygen.

The oxo reaction is not as simple as might be desired. Secondary reactions occur. For example, the aldehydes may undergo aldol condensation and the aldols may be hydrogenated to glycols. With alcohols and aldehydes in the same reactor, acetals may form. The net result is that up to 20% of the product may be high boiling.

While the oxo reaction with ethylene might at first glance seem to be the simplest oxo reaction, this is not the case for propionaldehyde forms less than half the reaction product. Evidently ethylene can add to propionaldehyde for almost half the product is diethyl ketone.

C. Iron Catalysts

There is a variation of the oxo process using $CO + H_2O$ instead of $CO + H_2$. The reaction with C_3H_6 can be represented:

$$C_3H_6 + 3\,CO + H_2O = C_4H_9OH + 2\,CO_2$$

This catalyst is also homogeneous and is based on iron pentacarbonyl (*11, 32*). The catalyst is made by reacting butylpyrolidine, water, and iron pentacarbonyl at 100°C. Propylene and carbon monoxide react in the aqueous solution at 100°C and 15 atm. The yield of *n*-butanol is reported to be 80–85% and of isobutanol 15–20%.

D. Poisons

Surprisingly, the cobalt carbonyl catalysts are remarkably insensitive to sulfur poisoning. Being acidic in nature, they are sensitive to bases.

The cobalt and nickel catalysts used for hydrogenating the aldehydes to alcohols are subject to CO and sulfur poisoning.

The writer does not have data on the poisons that may affect the Rh catalysts.

E. Activation of Second Stage Catalysts

Some of the catalysts supplied for the second stage must be reduced before use. One of the procedures given in Chapter 13 can be used. The following may be preferred by some. The catalyst is heated to 350°C in a circulating stream of nitrogen. Hydrogen is added to the circulating stream at a rate that will keep the maximum temperature in the bed below 400°C. When the gas stream is pure hydrogen and the temperature 400°C, keep the hydrogen flowing for 6 hours. Bring the catalyst down to reaction temperature and the pressure to operating pressure. The catalyst is active and pyrophoric.

F. Safety

Both cobalt and nickel catalysts which have been used for hydrogenating the aldehydes may be pyrophoric. The safety precautions and procedures given in Chapter 13 should be followed.

G. First Stage Catalysts

The first stage cobalt catalysts are made at the place of use. The following laboratory procedure may be useful. Cobalt acetate in methanol (or a cobalt salt of a higher molecular weight acid in a higher boiling alcohol, e.g., a product alcohol) in an autoclave is pressured to 150 atm with 1:1 H_2:CO and heated to 175°C for 1 hour. Maximum pressure is 230–250 atm. Cool the autoclave and the catalyst is ready to use.

To make the phosphine ligand catalyst, after cooling the autoclave, add the tributylphosphine or tricyclohexylphosphine and mix for 15 minutes at 40°C. This catalyst is then ready for use.

The following supply cobalt salts or chelates soluble in organic solvents:

Carlisle Chemical Works, Inc. (Also supply substituted phosphines.)
Harshaw Chemical Company
MacKenzie Chemical Works, Inc.
Nuodex Division of Tenneco Chemicals, Inc.
Witco Chemical Corporation, Organics Division.

Monochloro(carbonyl)bis(triphenylphosphine)rhodium(I), $((C_6H_5)_3P)_2Rh(CO)Cl$, is supplied by Matthey Bishop, Inc.

General Aniline and Film Corporation supplies iron pentacarbonyl.

H. Suppliers of Second Stage Catalysts

Catalysts Division of Chemetron Corporation

Girdler G–33: 33% Ni on kieselguhr, as ¼ inch tablets. The ABD is 1. Must be reduced before use.

Girdler G–60: a zirconium promoted nickel oxide on kieselguhr. Must be reduced before use.

Girdler G–60RS: the reduced and stabilized version of G–60.

Girdler G–61: about 60% Co, as oxide, on a refractory support. Must be reduced before use.

Girdler G–62: 33% Co, as oxide, on a refractory support. Must be reduced before use.

Girdler G–62RS: the reduced and stabilized form of G–62.

Girdler G–67: a zirconium promoted catalyst containing about 60% Co, as oxide, as $\frac{3}{16} \times \frac{1}{8}$ and $\frac{1}{4} \times \frac{3}{16}$ inch pellets. The ABD is 1.2. Must be reduced before use.

Girdler G–67RS: the reduced and stabilized form of G–67.

Harshaw Chemical Company

Co–0101 T: 35% Co, as oxide, on kieselguhr, as $\frac{3}{16}$ inch tablets. The ABD is 1.2, the surface area 130 m^2gm^{-1}, and the pore volume 0.35 cm^3gm^{-1}. Must be reduced before use.

Co–0108 T: 39% Co on kieselguhr, prereduced and stabilized, as $\frac{3}{16}$ inch tablets. The ABD is 1, the surface area 140 m^2gm^{-1}, and the pore volume 0.40 cm^3gm^{-1}.

Co–0403 G: 20% Co, as oxide, on a silica support, as 4 × 8 mesh granules. The ABD is 0.5, the surface area 6 m^2gm^{-1}, and the pore volume 0.38 cm^3gm^{-1}. Must be reduced before use.

Co–0501 T: 10% Co, as oxide, on activated alumina, as ⅛ inch tablets. The ABD is 1, the surface area 60 m^2gm^{-1}, and the pore volume 0.51 cm^3gm^{-1}.

Ni–0101 T: 44% Ni (as hydrate) on kieselguhr, as ⅛, $\frac{3}{16}$, and ¼ inch tablets. The ABD is 1.2–1.4. Must be reduced before use.

Ni–0104 T: 58% Ni on kieselguhr in which 60–65% of the total nickel has been reduced, as ⅛, $\frac{3}{16}$, and ¼ inch tablets. The ABD is 1.4–1.5, the surface area 125 m^2gm^{-1}, and the pore volume 0.2 cm^3gm^{-1}.

Universal Oil Products Company

H–2: 60% Ni on kieselguhr, as ⅛ inch extrudates. The ABD is 1.1.

H–4: 30% Ni on alumina, as ⅛ inch spheres. The ABD is 0.6–0.65.

II. Esterification

The reader is doubtless familiar with the use of hydrogen ions as esterification catalysts. It is old in chemistry to use such things as sulfuric acid or hydrogen chloride for this purpose. More recently, things like toluene sulfonic acid and similar sulfonic acids or BF_3 have been used. Also where these acids might cause darkening of the product, phosphorous acid or even hypophosphorous acid has been used.

Although the writer does not know of a commercial application, titania is a good catalyst where one can pass a mixture of the alcohol and the acid over a fixed bed of catalyst at a temperature in the range of 280–320°C. Conversion per pass is about 70%. The technique can be used to make the esters of formic acid, even though in the absence of the alcohol the formic acid would be decomposed. It can also be used with aromatic acids such as benzoic (*45*). The technique works well with primary alcohols, only fairly well with secondary alcohols, and is useless with tertiary alcohols.

Orthoesters of titanium are useful esterification catalysts in the liquid phase especially for secondary alcohols plus acids or phenols plus acids. Preferably, the water is removed as formed. Presumably, an active titanium catalyst is formed *in situ* from the ester.

Synthetic resins have been prepared with an open pore structure and the inside walls of the pores reacted with sulfonating agents to produce sulfonic acid groups and these can be used for esterification at temperatures up to 175°C. They provide a ready means for separating the acidic catalyst from the product.

III. Polyesters

It is often desirable to react a dihydroxy compound with a dibasic acid to form a high molecular weight polyester. Any catalyst used here is preferably left in the polyester, so it must not detract from the properties of the product. If the polyester contains unsaturation for later curing, the catalyst left in the polyester must not interfere with the curing or the inhibitors added to prevent premature curing. Obviously, the usual mineral acid catalysts do not meet these requirements. Further, it is not usually practical to pass the reacting mixture through a fixed bed of catalyst and obtain the high molecular weight products desired.

Orthoesters of titanium or zirconium, as well as titanium and zirconium chelates, have been found useful in quantities below 0.1% metal based on the charge. Apparently the zirconium catalysts tend to give lighter colored products and, unlike the titanates, they do not develop color with inhibitors.

The zirconium catalysts may or may not be the most active depending on the glycol–dibasic acid system (*51*).

3:1 stannous oxalate:sodium acetate alone or 3:1 stannous oxalate:sodium acetate + zinc acetate are also useful catalysts for forming polyesters.

A. Suppliers of Proven Catalysts

Anderson Chemical Division of Stauffer Chemical Company supplies tetrabutyl titanate, 2-ethylhexyl titanate and tetrabutyl zirconate.

Dow Chemical Company
Dowex 50: a cation exchange, synthetic resin in H^+ form.

E. I. du Pont de Nemours & Co., Inc. supplies a series of orthoesters and chelates of titanium under the trade name Tyzor.

MacKenzie Chemical Works, Inc. supplies titanium and zirconium chelates soluble in organic media.

Rohm and Haas Company
Amberlyst 15: a wide pore cation-exchange synthetic resin in which the walls of the pores have been sulfonated.

IV. Transesterification

One of the processes for forming a polyester is to condense an ester of a dibasic acid and a low molecular weight alcohol, e.g., methanol, with a dihydroxy compound so that the low boiling alcohol is liberated and distills off during the formation of the polyester. The orthoesters of titanium or zirconium described above are useful for this purpose. Or small proportions of antimony trioxide may be used. Mixtures containing antimony trioxide, e.g., 0.035% ZnO + 0.1% Sb_2O_3 based on the ester–glycol, have been used as have organo tin compounds. For example, 0.1% to 1% dibutyl tin dilaurate, butyl hydroxystannic oxide or dibutyltin oxide are catalysts. Antimony trifluoride can be used in place of antimony oxide.

The reaction is ordinarily carried out in a batch reactor at 190°–200°C, first at atmospheric pressure, then at lower pressure as the mixture becomes more viscous. The final pressure is around 0.05 Torr and the final temperature around 225°C. The catalyst is left in the polyester.

In a different version of this process, low boiling esters of fat acids, e.g., methyl esters, may be made by heating dry glycerides and an excess of methanol (or other alcohol) with 0.1–1% of the sodium alkoxide as the catalyst.

A. Suppliers of Proven Catalysts

The suppliers of the titanium and zirconium catalysts are given on page 223.

Carlisle Chemical Works, Inc.
Carstan 8: stannous octoate.
Carstan 18: stannous oleate.
Carstan DBTDL: dibutyltin dilaurate.

M & T Chemicals, Inc.
Dibutyltin oxide, dibutyltin diacetate, dibutyltin maleate, dibutyltin dilaurate as well as the following:
T–8: dibutyltin di-2-ethylhexoate.
T–18: the stannous salt of a long chain fat acid.

Nuodex Division of Tenneco Chemicals, Inc.
Nuocure 28: tin octoate.

V. Interesterification

This is the process of randomizing fat acids in glycerides. It improves the shortening properties of cooking fats, especially lard for baking.

The reaction is carried out at 55–95°C with 0.2% sodium methoxide as the catalyst. The fat must be dry. The color changes from light brown to dark brown as the reaction reaches equilibrium, and there is no further lowering of the melting point. The catalyst is destroyed by adding water and the soap formed is removed. Any methyl esters formed are removed during stripping with steam to deodorize.

Alternatively, orthoesters and chelates of titanium may be used as catalysts at the same or slightly higher temperature. About 1% catalyst is dissolved in the glyceride. For food purposes the titanium must be removed.

VI. Vinyl Acetate from Acetylene

For many years, acetylene + acetic acid was the primary route to vinyl acetate. The less expensive ethylene is rapidly replacing acetylene as a source of vinyl acetate (see Chapter 20). Broadly, for obtaining vinyl acetate from acetylene, there is a liquid phase and a vapor phase process (*14*).

In the liquid phase process, C_2H_2 is passed into a solution of the mercury catalyst at 60°–80°C. With a large excess of C_2H_2, the vinyl acetate leaves the reactor as vapor and is condensed. The catalyst is prepared by reacting HgO with enough oleum in acetic acid to provide 1.5% excess H_2SO_4. The concentration of catalyst (as HgO) is 2.5% in the acetic acid. As the acetic acid is consumed in the reaction, more is added to maintain the catalyst concentration. The reaction is strongly exothermic so cooling is essential. With use the catalyst is reduced to Hg and must be reworked.

In the vapor phase process, 3 moles of acetylene + 1 mole of acetic acid are passed over a cadmium, zinc, or mercury on charcoal catalyst at 170°–225°C. It is preferable to limit the catalyst temperature to 170°C with fresh catalyst and raise the temperature as the catalyst ages to maintain the production rate. Cooling must be provided to control the temperature.

A. Suppliers of Proven Catalysts

Calsicat Division of Mallinckrodt Chemical Works supplies mercuric chloride on carbon and zinc acetate on carbon.

Harshaw Chemical Company
- C–0301 G: 10% mercuric chloride mounted on activated carbon, as 4 × 10 mesh granules.
- Zn–0803 G: 10% Zn as acetate on carbon, as 4 × 10 mesh granules. The ABD is 0.7.

VII. Acrylic Esters from Acetylene

Nickel carbonyl is a catalyst for a process that in its simplest form can be represented:

$$C_2H_2 + CO + ROH = H_2C{=}CHCOOR$$

The process, which is carried out at 30–60°C is more complicated. HCl and $Ni(CO)_4$ in a ratio of 1.01:1–1:1.2, C_2H_2:CO of 1.01:1–1.1:1, and an alcohol:CO ratio of 1.1:1–3:1 are all fed simultaneously and continuously in closely controlled proportions to the reactor. About 80% of the necessary CO comes from CO in the feed and 20% from CO in the $Ni(CO)_4$. Thus $Ni(CO)_4$ serves both as a reagent and a catalyst (*6*, *46*).

An excellent study is available showing the effect of process variables, detailing kinetics, and proposing a mechanism (*19*). Interestingly, under the conditions studied, CO was a poison when present in excess of that required by the C_2H_2. Apparently the process is not readily adaptable to the higher acetylenes.

A catalyst made from $NiBr_2$ and triphenylphosphine in alcoholic solution gives 70–80% yields of acrylic esters from CO + C_2H_2 at 150°–160°C provided some alkyl bromide is present (*27*).

Tetrahydrofuran is a useful solvent for acetylene when it is to be introduced into reactions under high pressure (*12*).

A. Safety

Acetylene without a solvent, under high pressure, is hazardous. Safety manuals should be consulted.

Nickel carbonyl is both spontaneously flammable and poisonous. It is usually made where it is needed when used commercially.

VIII. Polyurethanes

Aryl isocyanates combine with alcohols to give urethanes, esters of carbamic acids. Using diisocyanates and diols, flexible polyurethanes may be obtained. With triols or polyols, rigid products are made. By including some water in the mixture, this reacts with the isocyanate to produce a carbamic acid which decomposes to an amine and CO_2. The CO_2 liberated during the reaction causes the reacting mixture to "rise" or "foam." Like in baking bread, the objective is to have the condensation reaction well along before the CO_2 starts forming so it does not escape and let the foam collapse. Catalysts are added to speed up and control the rates of the reactions. There is a summary of the wide variety of catalysts that have been examined and show catalytic activity (*21*). The more effective ones are tertiary aliphatic amines and two common ones are *N*-ethylmorpholine and 1,4-diazabicyclo-(2.2.2)-octane. Many metal compounds soluble in the reaction mixture are also catalysts, with tin compounds being outstanding (*21*). Practically, these two types of catalysts are synergistic, so when fast reaction is wanted (which is usually the case) the two catalyst types are used together. The concentration and ratio of the catalysts depends on the system but the range is usually 0.05–0.5% by weight of each tin compound and amine. Although some heat is evolved in the process, the reaction is carried out at substantially room temperature. The components are mixed at a mixing nozzle and the mixture flows into a mold where the reaction is completed in a matter of minutes.

A. Suppliers of Proven Catalysts

Carlisle Chemical Works, Inc.
Carstan 8: stannous octoate.
Carstan 18: stannous oleate.
Carstan DBTDL: dibutyltin dilaurate.

Houdry Process and Chemical Company
Dabco: 1,4-diazabicyclo-(2.2.2)-octane.

M & T Chemicals, Inc.
Dibutyltin dilaurate.
T–8: dibutyltin di-2-ethylhexoate.
T–18: the stannous salt of a long chain fat acid.

Nuodex Division of Tenneco Chemicals, Inc.
Nuocure 28: tin octoate.

IX. 2-Methyl-5-ethylpyridine

Acetaldehyde is polymerized to paraldehyde by a trace of sulfuric acid. This is mixed with aqua ammonia (30–40% NH_3) and a little acetic acid (ammonium acetate is the catalyst) and reacted continuously at 220°–280°C and 100–200 atm pressure. The reaction takes place in the liquid phase. The reaction product separates into two layers on cooling. Most of the aqueous layer is fortified with NH_3 and recycled. The 2-methyl-5-ethylpyridine is recovered from the organic layer. The yield is about 70% of theory based on acetaldehyde (*9*, *36*).

It should be noted that 2-methyl-5-ethylpyridine can be dehydrogenated to 2-methyl-5-vinylpyridine using the same catalysts and essentially the same conditions used for converting ethylbenzene to styrene (Section III, Chapter 6).

X. Aliphatic Amines

The interaction of alcohols with ammonia to produce alkylamines in the presence of dehydration catalysts can be regarded as one of the classical Sabatier reactions (*43*). The reaction has several characteristics.

(1) It produces a mixture of primary, secondary, and tertiary amines.

(2) The ratio of products can be controlled to a limited extent by controlling the ratio of the reactants.

(3) The ratio of products can also be controlled to some extent by recycling unwanted amines.

(4) The reaction works best with methanol. Higher alcohols are dehydrated to some extent to olefinic hydrocarbons.

(5) The reaction works best with primary alcohols, only fairly well with secondary alcohols, and is useless with tertiary alcohols.

(6) The reaction does not give useful yields with phenols, but arylcarbinols, e.g., benzyl alcohol, will work.

Reaction conditions are 300°–400°C, substantially atmospheric pressure, and a LHSV (based on the alcohol) of 0.5–5 $V_l V_c^{-1}$ hr^{-1}. Operating pressure can be higher, e.g., 5 atm, when this is convenient in processing the reaction products.

Since the reaction is capable of making secondary and tertiary amines, it is also capable of making mixed amines. For example, methylaniline and dimethylaniline can be made from aniline and methanol or methylaniline and methanol (*38*). Both silica–alumina and alumina catalysts were used but the alumina tended to give more dimethyl ether in converting methylaniline to dimethylaniline.

The reaction is applicable to higher molecular weight aliphatic or cycloaliphatic alcohols and lower molecular weight amines, e.g., dodecyl alcohol plus dimethylamine (large excess) to give dimethyldodecylamine.

Many different dehydration catalysts have been used and the reader is referred to Chapter 5 for a discussion of the catalysts and a list of suppliers.

A. Supplier of Proven Catalyst

Davison Chemical Division of W. R. Grace & Co.
Grade 970: a specially modified silica–alumina for making methylamines.

XI. Alkyl Mercaptans and Sulfides

Like the preceding reaction, this is one of the Sabatier classics (*44*) in which an alcohol and hydrogen sulfide react to give mercaptans and sulfides. The reaction has characteristics analogous to those listed for the alkylamines above.

Reaction conditions are 250°–400°C, atmospheric pressure, an LHSV (based on alcohol) of 0.3–3 $V_l V_c^{-1}$ hr^{-1}, and a large excess of H_2S (3–7 moles per mole of alcohol) for the production of the mercaptan (*24, 25*). Neutral rather than acidic γ-alumina or γ-alumina to which 5–10% K_2CO_3 or K_2WO_4 has been added are preferred catalysts.

Alkyl mercaptans can also be made by adding H_2S to olefinic hydrocarbons. The reaction takes place with acidic catalysts, e.g., anhydrous $AlCl_3$ at 0°C or supported phosphoric acid at 80°C. In contrast to the high temperature alcohol + H_2S reaction which works best with primary alcohols, this process works best to make tertiary alkyl mercaptans. Some dialkyl sulfide is formed by the addition of the mercaptan to the olefin.

H_2S may also add to olefins in the presence of the cobalt–molybdenum oxide–alumina type catalysts to give mercaptans (see Chapter 16 for suppliers). The conditions reported for $C_2H_4 + H_2S$ are 280°–300°C, 8 atm pressure, and $H_2S:C_2H_4$ ratio 4:1 (*1*). Other gases such as methane or hydrogen were present in the reaction mixture.

Using free radical initiators such as peroxides, hydroperoxides, etc., H_2S adds anti-Markownikoff to olefins.

Aldehydes, ketones, and nitriles can be converted to mercaptans by sulfactive catalysts, e.g., cobalt sulfide, molybdenum sulfide, and nickel sulfide. The starting material, a solvent and sulfur are heated together with hydrogen at 300°–400°C and 30–70 atm pressure (*22*). In a batch process, the amount of catalyst is 5–15% of the starting material.

The writer does not know of a supplier for the catalysts mentioned. If he had a problem of this nature and preferred to use a commercial catalyst he would use one of those of the cobalt–molybdenum oxide–alumina type described in Chapter 16.

A. Safety

H_2S is not only poisonous, it is highly corrosive to most ferrous metals. Reactors used under pressure with H_2S should be reexamined frequently to be certain they are still safe to use.

XII. Ethanol to Butadiene

During World War II, large quantities of butadiene for synthetic rubber were manufactured from ethanol. A part of the ethanol was dehydrogenated catalytically to acetaldehyde and this condensed in the presence of ethanol and a tantala on silica gel catalyst to produce butadiene. Typical operating conditions are 325°C, a ratio of 3 moles of ethanol to 1 of acetaldehyde, and LHSV 0.33–0.5 $V_l V_c^{-1}$ hr^{-1}, all at atmospheric pressure, using 2% tantala deposited on silica gel. Two percent zirconia on silica gel is almost as good (*55*).

As might be expected, a tremendous amount of work was expended to see if the preceding two-step process of dehydrogenation followed by dehydration could be converted into a single-step process using a combination catalyst that would do both the dehydrogenation and dehydration *in situ*. It was possible to do this although the efficiency of the utilization of ethanol was not quite as good. For example, a catalyst made of 59% MgO, 2% Cr_2O_3, and 39% silica or 60% MgO, 2% tantala, and 38% silica gave satisfactory results at 425°C and LHSV 0.4 $V_l V_c^{-1}$ hr^{-1} and atmospheric pressure (*17*). The author is not aware of a commercial supplier of either of these two catalysts.

A. Supplier of Proven Catalyst

Davison Chemical Division of W. R. Grace & Co.
Grade 927: 2% tantala on silica gel.

XIII. Ethanol Plus Crotonaldehyde to Butadiene

In developing the alcohol to butadiene process, it was found that ethanol + crotonaldehyde could be reacted to produce butadiene + acetaldehyde. Silica gel is a usable catalyst and 2% tantala on silica gel is even better (see Section XII,A for supplier). The reaction conditions for the latter catalyst are 325°C, LHSV 0.7 $V_l V_c^{-1}$ hr^{-1}, C_2H_5OH:crotonaldehyde mole ratio 6:1. Conversion per pass is 90+% (*55*).

In this process, the ethanol presumably transfers hydrogen to the crotonaldehyde to form crotyl alcohol. This is then dehydrated to butadiene. Conceivably the process could be applied to other systems.

XIV. Urea to Melamine

Urea is decomposed into isocyanic acid and ammonia in a heated fluidized solids reactor. The solids are noncatalytic. The isocyanic acid is dehydrated and cyclicized to melamine by a dehydration catalyst at 400°–430°C (*47*). See Chapter 5 for catalysts and suppliers.

XV. Hydration of Olefins to Alcohols

The hydration of olefinic hydrocarbons is probably the most common hydration process. This is used to produce ethanol from ethylene, isopropanol from propylene, and the butanols from butylenes. For example, isopropanol is made by absorbing liquid or gaseous propylene in 75% sulfuric acid at room temperature. The product, diluted with water, is hydrolyzed with steam and the isopropanol removed with steam (*4*). The reader may reason that the sulfuric acid is actually a reagent, that an intermediate, isopropylsulfuric acid, is formed and then hydrolyzed. The reader is right, and similar intermediates are formed and react in other catalytic reactions. As in other catalytic reactions, equilibria are involved. Isopropanol is present in the sulfuric acid–isopropylsulfuric acid before dilution with water and may be stripped out with another gas. Adding water to replace that consumed in the hydration of the propylene aids in stripping the isopropanol. In most catalytic reactions, neither the intermediate compound nor the equilibria are so easily identified and followed.

Using phosphoric acid instead of sulfuric avoids the undesirable oxidizing properties of sulfuric acid. For fixed bed catalysts, phosphoric acid on kieselguhr or silica gel or acid phosphates are useful. For example, one process uses phosphoric acid on kieselguhr to produce ethanol from ethylene

(*13*). Temperature is 300°C, pressure 70 atm with 0.6 moles of H_2O per mole of C_2H_4. The ethylene conversion per pass is 4.2% and water conversion 7%. Conditions for fixed bed operation with propylene are 180°–260°C and 25–65 atm (*10*).

A. Suppliers of Proven Catalysts

Catalysts and Chemicals, Inc.
C84: phosphoric acid on kieselguhr as 5/16–3/8 inch spheres.

Davison Chemical Division of W. R. Grace & Co.
Grade 57: silica gel specially developed as a carrier for phosphoric acid for hydrating ethylene to ethanol.

Universal Oil Products Co.
Phosphoric acid on kieselguhr, No. 1, 1/4 inch; No. 2, 3/16 inch; and No. 4, 3/16 inch extrudates.

Care must be used with the supported phosphoric acid catalysts. Too much water in the feed can leach acid out of the catalyst. With phosphoric acid on kieselguhr, it can soften the catalyst particles so they collapse and increase pressure drop in the bed.

XVI. Saponification

Both acids and bases are used to hydrolyze esters. The manufacture of soap from fats and oils using sodium hydroxide as both catalyst and reagent is the commonest process. Possibly the most common acid catalyst process is the Twitchell process for hydrolyzing fats to fat acids and glycerine. In this process, the fat plus 25–50% water, 0.75–1.25% Twitchell catalyst, and 0.5% sulfuric acid are boiled together for 20–48 hours. The glycerine formed dissolves in the excess water and is separated from the molten fat acids (*34*).

The Twitchell catalyst is an alkylaryl sulfonic acid whose organic part is large enough to make it soluble in the fat. Mahogany acids, formed in the sulfuric acid treating of petroleum oils, are frequently used in place of Twitchell catalyst. The mahogany acids are also sulfonic acids whose organic part is large enough to make them soluble in the fat.

A. Catalyst Supplier

Witco Chemical Company, Sonneborn Division, supplies mahogany sulfonates.

XVII. Hydrolysis of Chlorobenzene to Phenol

In the Rashig process for making phenol, chlorobenzene is hydrolyzed to phenol in the presence of a silica gel catalyst at about 500°C and substantially atmospheric pressure (*7*, *37*). Conversion per pass is about 10%.

A. Supplier of Proven Catalysts

Davison Chemical Division of W. R. Grace & Co.

RD–Gel: in particle sizes from 3 × 8 mesh to <200 mesh. The ABD is 0.7, the surface area 800 m^2gm^{-1}, and the pore volume 0.45 cm^3gm^{-1}.

ID–Gel: as >10 mesh granules. The ABD is 0.45, the surface area 350 m^2gm^{-1}, and the pore volume 1.04 cm^3gm^{-1}.

Grade 02: as >4 mesh granules. The ABD is 0.7, the surface area 525 m^2gm^{-1}, and the pore volume 0.30 cm^3gm^{-1}.

XVIII. Hydration of Acetylene

The hydration of acetylene to acetaldehyde using acid activated by mercury salts, e.g., mercuric bisulfate, is a relatively old process (*29*). It is being rapidly replaced by the catalytic oxidation of ethylene to acetaldehyde (see Chapter 19). The acetylene process is conducted at 50°–100°C, and the acetaldehyde is carried out of the reaction mixture by excess acetylene and recovered. The reaction is strongly exothermic.

XIX. Chlorination

The addition of chlorine to the carbon–carbon double bond is not as easy as some textbooks imply. Low temperatures, e.g., −25°C, favor the dichloride. At higher temperatures substitution becomes a factor. The presence of some liquid $C_2H_4Cl_2$ aids the reaction. Catalysts are supported copper chloride. The reaction is highly exothermic and should be conducted at the lowest practical temperature the activity of the catalyst and heat removal facilities permit. The reaction can be carried out in a fluidized solids reactor or in a fixed bed.

A. Suppliers of Proven Catalysts

Harshaw Chemical Company

Cu–0905 P: 10% $CuCl_2$ mounted on activated alumina, as a powder. The ABD is 1 and the surface area 180 m^2gm^{-1}.

Cu–0905 T: 10% $CuCl_2$ mounted on activated alumina, as ⅛ inch tablets. The ABD is 1, the surface area 59 m^2gm^{-1}, and the pore volume 0.26 cm^3gm^{-1}.

XX. Hydrochlorination and Hydrofluorination

Bismuth trichloride or antimony trichloride at 120°–200°C are catalysts for the addition of hydrogen chloride to ethylene. These catalysts will catalyze the addition of hydrogen chloride to propylene at room temperature (*35*).

Activated charcoal itself is a moderately active catalyst for this process, and if zinc chloride or anhydrous stannic chloride is supported on the activated charcoal, it becomes still more active. Tertiary alkyl halides are so unstable that it is difficult to form them at temperatures above about 70°–80°C. For this reason, it is possible to separate the secondary and the tertiary alkyl halides by operating the hydrohalogenation at 150°C to produce the secondary halides, and then to separate these before conducting the reaction at 70°C to form the tertiary chloride (*40*).

HF can be added to olefins using sodium fluoride, aluminum fluoride, or chromium catalysts that presumably are converted to fluorides under reaction conditions.

A. Suppliers of Proven Catalysts

Harshaw Chemical Company

Cu–0905 T: 10% $CuCl_2$ mounted on activated alumina, as ⅛ inch tablets for the hydrochlorination of olefins. The ABD is 1, the surface area 59 m^2gm^{-1}, and the pore volume 0.26 cm^3gm^{-1}.

Al–1101 T: 85% aluminum fluoride, as ¼ inch tablets. The ABD is 1.3.

Cr–0304 T: 33% Cr_2O_3 on alumina, as ⅛ inch tablets. The ABD is 1.65, the surface area 120 m^2gm^{-1}, and the pore volume 0.26 cm^3gm^{-1}.

Na–0202 T: sodium fluoride, as ⅛ inch tablets. The ABD is 1 and the surface area 1 m^2gm^{-1}.

XXI. Hydrohalogenation of Acetylene

The addition of HCl to acetylene gives vinyl chloride. A mixture of HCl and C_2H_2 is passed over a catalyst bed of mercuric chloride deposited on carbon. The reaction is exothermic so that adequate cooling must be provided to keep the temperature in the range of 160°–250°C.

Laboratory data are available on various silica supported metal halides as catalysts (*50*).

A. Suppliers of Proven Catalysts

Calsicat Division of Mallinckrodt Chemical works supplies mercuric chloride on carbon.

Harshaw Chemical Company

C–0301 G: 10% mercuric chloride mounted on activated carbon, as 4 × 10 mesh granules.

XXII. Acrylonitrile from Acetylene

While obsolescent, the addition of HCN to C_2H_2 to make acrylonitrile has been an important commercial process. The catalyst is a 36% solution of Cu_2Cl_2, HCl, and alkali chlorides in water. The reactor is a rubber lined cylinder containing catalyst solution to a depth of a meter or two. The reaction temperature is 80°C. About 15% of the C_2H_2 and all the HCN introduced are consumed (*5*).

XXIII. Dehydrochlorination

Many alkyl halides are unstable enough so that they will decompose into an olefin and HCl just by heating. This process is important in several commercial areas. In its simplest form, ethylene dichloride is heated to convert it into vinyl chloride + HCl. In a similar way, 1,1,2-trichloroethane is pyrolyzed to convert it into vinylidene chloride and hydrogen chloride. Carrying the process one step further, tetrachloroethane (acetylene tetrachloride) is heated to convert it into trichloroethylene and hydrogen chloride.

If there is any need to lower the temperature of the reaction from the thermal reaction temperatures of 500°–550°C, this can be done with catalysts. Activated charcoal has some activity but barium chloride deposited on carbon is usually employed. The temperature depends on the material to be dehydrohalogenated but around 300°C is a good place to start. Mercuric chloride deposited on carbon has also been used.

Sometimes halogen can be removed by hydrogenation. For example, chlorotetracycline is converted to tetracycline + hydrogen chloride by hydrogen in the presence of a finely divided Pd catalyst.

A. Suppliers of Proven Catalysts

See Chapter 13 for the suppliers of finely divided Pd catalysts on several supports.

Harshaw Chemical Company

Ba–0108 E: 27% $BaCl_2$ on carbon, as 4 × 6 mesh extruded particles. The ABD is 0.5.

C–301 G: 10% $HgCl_2$ on activated carbon, as 4 × 10 mesh granules.

XXIV. Removing Traces of Halides

Frequently, it is desirable to remove relatively small quantities of halides from a feed material, e.g., one that is to be hydrogenated by a nickel catalyst that is poisoned by halogens. Raney nickel (activated) is an active reagent for reacting with such halogen and removing it. Obviously, the Raney nickel becomes deactivated in the process so that it is not suitable for hydrogenation purposes. In some cases, it will be possible to use a nickel catalyst that is relatively spent for hydrogenation as a "guard case" for removing halogen and other catalyst poisons before a stream goes into the main catalyst bed.

XXV. Dehydrofluorination

In the HF alkylation of isobutane with olefins, some alkyl fluorides are formed which are not wanted in the product. In addition, HF is soluble to about 1% in the alkylate–butane product, and this must be recovered. A "propane" stripper is used to recover this HF and this is frequently packed with aluminum metal Raschig rings to defluorinate at least partly the alkyl fluorides. This catalytic reaction is reversible:

$$RCH_2{-}CH_2F \rightleftharpoons RCH{=}CH_2 + HF$$

This means that the olefin and HF must be separated or they may recombine. This has occasionally led to a misunderstanding of this combination stripper–catalyst reactor. It has been erroneously concluded that the catalyst was not doing any good when in fact it was being operated so that the products were reacting to reform the original fluorides.

The dehydrofluorination with aluminum Raschig rings takes place in the 30°–100°C temperature range at 15–20 atm pressure. Typical operating conditions are 15 atm and 60°C overhead temperature in the stripper.

It is not practical to decompose all alkyl fluorides in the catalyst–stripper. A low-silica bauxite is used to decompose the rest. The bauxite actually combines with the liberated HF so that the bauxite must be replaced when spent. The operating temperature is 100°–110°C.

XXVI. Fluorohydrocarbons from Chlorohydrocarbons

This reaction can be represented by the equation:

$$CCl_4 + x\,HF = CCl_{4-x}F_x + x\,HCl$$

Catalysts are the trihalides of Sb, Al, or Cr, which may be converted to the fluorides if they are not fluorides at the start of the reaction (*8*, *57*).

XXVII. Fischer–Tropsch Process

This process has been described in detail (*2*, *3*, *15*, *39*, *52–54*, *58*). It has had a colorful history in that it was developed in Germany in the late 1920's and operated there commercially through World War II. There the process was based on a nickel on kieselguhr catalyst promoted by thorium and potassium compounds used in fixed beds. After World War II, there was a flurry of activity in the U.S. on the process as the nation feared a shortage of petroleum. The process was based on synthesis gas from natural gas instead of coal as the Germans had used. Two plants were designed and one was built based on an iron catalyst similar to the synthetic ammonia catalyst operated in a fluidized bed. The plant was shut down for economic reasons. As of this writing, the author knows of only one plant operating and it is based on coal. It employs five fixed bed reactors operating at 220°–225°C and three fluidized solids reactors operating at about 320°C. An indication of the product distributions is given in Table XVIII (*41*, *42*). The production

TABLE XVIII

PRODUCT DISTRIBUTION FROM FISCHER–TROPSCH REACTORS[a]

	Fluidized catalyst	Fixed bed catalyst
LPG ($C_3 + C_4$)	7.7%	5.6%
Gasoline (C_5–C_{12})	72.3%	33.4%
Gas oil (diesel fuel)	3.4%	16.5%
Crude paraffin (waxy oil)	3.0%	10.3%
Soft paraffin (MP 57°–60°C)	—	11.8%
Hard paraffin (MP 95°–97°C)	—	18.0%
Alcohols and ketones	12.6%	4.3%
Organic acids	1.0%	Trace

[a] See references (*41*) and (*42*)

TABLE XIX

PRODUCTION RATES FROM FLUIDIZED SOLIDS FISCHER–TROPSCH REACTORS[a]

Product	Production rate (m^3 day^{-1})
LPG	3.8
Gasoline	520
Diesel fuel	22.6
Waxy oil	7.3
Methanol	2.3
Ethanol (crude)	1.9
Ethanol (pure anhydrous)	48
Methyl ethyl ketone	3.5
Acetone	2.5

[a] See reference (*26*).

rates of the three fluidized solids reactors are summarized in Table XIX (*26*).

Operating conditions for the fluidized catalyst system have been given: Temperature 310°–345°C, pressure 17–25 atm, H_2:CO 2.8:1–3.3:1 and gas recycle ratio of 2:1–2.5:1. The linear velocity in the reactor is 1.3–2.3 m sec^{-1}.

The catalysts are prepared at the plant site by the electrothermal fusion of high grade magnetite with the controlled addition of alkali promoters (mostly potassium compounds). For the fluidized bed, the 40 μ powder is activated by reducing it in a separate reactor with hydrogen at 17 atm and 370°–400°C. This is then conditioned at lower pressures and high H_2:CO ratio gas to generate a combination of Fe_2C, Fe_3O_4, and Fe that is the catalyst. The composition changes with use.

A. Supplier of Catalyst

The author is not aware of a supplier of this catalyst. If he had a problem in this area, he would start with a synthetic ammonia catalyst and reduce it under the conditions given above. See Chapter 14 for suppliers of synthetic ammonia catalysts.

REFERENCES

1. Allison, J. A. C., and Benvegno, G. J., *Am. Chem. Soc. Div. Petrol. Chem. Preprints* **13,** No. 3, 207 (1968).
2. Anderson, R. B., *Advan. Catalysis* **5,** 355 (1953).
3. Anderson, R. B., *Catalysis* **4,** 1, 29, 257 (1956).
4. Anonymous, *Hydrocarbon Process. Petrol. Refiner* **42,** No. 11, 189 (1963).
5. Anonymous, *Hydrocarbon Process. Petrol. Refiner* **42,** No. 11, 138 (1963).
6. Anonymous, *Hydrocarbon Process. Petrol. Refiner* **44,** No. 11, 242 (1965).
7. Anonymous, *Hydrocarbon Process. Petrol. Refiner* **44,** No. 11, 256 (1965).
8. Anonymous, *Hydrocarbon Process. Petrol. Refiner* **44,** No. 11, 214 (1965).
9. Anonymous, *Hydrocarbon Process. Petrol. Refiner* **44,** No. 11, 243 (1965).
10. Anonymous, *Hydrocarbon Process.* **46,** No. 11, 195 (1967).
11. Anonymous, *Hydrocarbon Process.* **46,** No. 11, 153 (1967).
12. Anonymous, *Hydrocarbon Process.* **46,** No. 11, 140 (1967).
13. Anonymous, *Hydrocarbon Process.* **46,** No. 11, 168 (1967).
14. Argenbright, R. P., and Evans, R. J., *Hydrocarbon Process. Petrol. Refiner* **43,** No. 11, 159 (1964).
15. Arnold, J. H., and Keith, P. C., *Advan. Chem. Ser.* **5,** 120 (1951).
16. Asinger, F., Fell, B., and Rupilius, W., *Ind. Eng. Chem. Prod. Res. Develop.* **8,** 214 (1969).
17. Corson, B. B., Jones, H. E., Welling, C. E., Hinckley, J. A., and Stahly, E. E., *Ind. Eng. Chem.* **42,** 359 (1950).
18. Craddock, J. H., Hershman, A., Paulik, F. E., Robinson, K. K., and Roth, J. F., *Ind. Eng. Chem. Prod. Res. Develop.* **8,** 291 (1969).
19. Ehrreich, J. E., Nickerson, R. G., and Ziegler, C. E., *Ind. Eng. Chem. Process Design Develop.* **4,** 77 (1965).
20. Evans, D., Osborn, J. A., and Wilkinson, G., *J. Chem. Soc.* **A,** 3133 (1968).
21. Farkas, A., and Mills, G. A., *Advan. Catalysis* **13,** 393 (1962).
22. Farlow, M. W., Lazíer, W. A., and Signaigo, F. K., *Ind. Eng. Chem.* **42,** 2547 (1950).
23. Fell, B., Rupilius, W., and Asinger, F., *Tetrahedron Letters* p. 3261 (1968).
24. Folkins, H. O., and Miller, E. L., *Ind. Eng. Chem. Process Design Develop.* **1,** 27 (1962).
25. Folkins, H. O., and Miller, E. L., *Proc. Am. Petrol. Inst., Sect. III* **42,** 188 (1962).
26. Garrett, L. W., Jr., *Chem. Eng. Progr.* **56,** No. 4, 42 (1960).
27. Germain, J. E., "Catalytic Conversion of Hydrocarbons," p. 286. Academic Press, New York, 1969.
28. Hershman, A., and Craddock, J. H., *Ind. Eng. Chem. Prod. Res. Develop.* **7,** 226 (1968).
29. Hilditch, T. P., and Hall, C. C., "Catalytic Processes in Applied Chemistry," 2nd Ed., p. 261. Van Nostrand, New York, 1937.
30. Knap, J. E., Cox, N. R., and Privette, W. R., *Chem. Eng. Progr.* **62,** No. 4, 74 (1966).
31. Kobe, K. A., and McKetta, J. J., Jr., *Advan. Petrol. Chem. Refining* **1,** 486 (1958).
32. von Kutepow, N., and Kindler, H., *Angew. Chem. Intern. Ed. English* 41 (1961).
33. Lempke, H., *Hydrocarbon Process.* **45,** No. 2, 148 (1966).
34. Marsel, C. J., and Allen, H. D., *Chem. Eng.* **54,** No. 6, 104 (1948).
35. Maxted, E. B., "Catalysis and Its Industrial Applications," p. 387. Churchill, London, 1933.
36. Nenz, A., and Pieroni, M., *Hydrocarbon Process.* **47,** No. 11, 139 (1968); **47,** No. 12, 103 (1968).
37. Olive, T. R., *Chem. Met. Eng.* **47,** 770 (1940).
38. Parera, J. M., Gonzalez, A., and Barral, M. A., *Ind. Eng. Chem. Prod. Res. Develop.* **7,** 259 (1968).

39. Pichler, H., *Advan. Catalysis* **4,** 272 (1952).
40. Piotrowski, W. J., and Winkler, J., *J. Inst. Petrol. Technologists* **17,** 225 (1931).
41. von Rousseau, P. E., *Chem. Ind.* (*London*) p. 1958 (1962).
42. von Rousseau, P. E., van der Merve, J. W., and Louw, J. D., *Brennstoff-Chem.* **44,** No. 6, 166 (1963).
43. Sabatier, P., and Mailhe, A., *Compt. Rend.* **150,** 823 (1910).
44. Sabatier, P., and Mailhe, A., *Compt. Rend.* **150,** 1570 (1910).
45. Sabatier, P., and Mailhe, A., *Compt. Rend.* **152,** 494 (1911).
46. Salkind, M., Riddle, E. H., and Keefer, R. W., *Ind. Eng. Chem.* **51,** 1328 (1959).
47. Schmidt, A., *Hydrocarbon Process.* **45,** No. 11, 146 (1966).
48. Slaugh, L. H., and Mullineaux, R. D., U.S. Patent 3,239,569 (March 8, 1966).
49. Slaugh, L. H., and Mullineaux, R. D., U.S. Patent 3,239,566 (March 8, 1966).
50. Smith, D. M., Walsh, P. M., and Slager, T. L., *J. Catalysis* **11,** 113 (1968).
51. Stevens, M. P., and Gardener, J. D., *Ind. Eng. Chem. Process Design Develop.* **4,** 67 (1965).
52. Storch, H. H., *Advan. Catalysis* **1,** 115 (1948).
53. Storch, H. H., *Advan. Chem. Ser.* **5,** 120 (1951).
54. Storch, H. H., Golumbic, N., and Anderson, R. B., "The Fischer-Tropsch and Related Syntheses." Wiley, New York, 1951.
55. Toussaint, W. J., Dunn, J. T., and Jackson, D. R., *Ind. Eng. Chem.* **39,** 120 (1947).
56. Tucci, E. R., *Ind. Eng. Chem. Prod. Res. Develop.* **7,** 32, 227 (1968); **8,** 215 (1969).
57. Vecchio, M., Cammarata, I., and Cevidalli, G., *Chem. Eng. Progr.* **60,** No. 9, 51 (1964).
58. Weil, B. H., and Lane, J. C., "Synthetic Petroleum from the Synthine Process." Chem. Publ. Co., New York, 1948.
59. Wender, I., Sternberg, H. W., and Orchin, M., *Catalysis* **5,** 73 (1958).

CHAPTER 22

CATALYST CARRIERS AND SUPPORTS

Catalyst carriers were originally employed to spread out or disperse a catalytically active material, e.g., Pt, so that it could be used more effectively. It was presumed that the carrier was inert and merely supported the active material. While the carrier may be inert as far as the reaction is concerned, most of them retard or prevent the crystallization of the catalyst and so extend its useful life. Section II of this chapter discusses these inert carriers.

Some carriers are not inert and, while giving the appearance of being a support, actually form a vital part of the catalyst. For example, chromia–alumina might seem to be a chromia catalyst on an alumina carrier. However, chromia on other carriers does not provide the catalytic properties found in chromia–alumina. A similar situation is found in the cobalt oxide–molybdenum oxide–alumina catalysts, and there are many others.

The distinction between catalyst and carrier becomes even more diffuse in the zeolite cracking catalysts where 10–15% of highly active ion-exchanged zeolite is dispersed in an amorphous silica–alumina matrix that is also a cracking catalyst. If Pd, for example, is supported on the composite catalyst to produce hydrocracking catalysts, the system has a *prima facie* appearance of a supported catalyst, but all components are active and contributing to the overall reaction.

In a different sense, some materials which are carriers for catalysts are catalysts in themselves for different reactions. For example, γ-alumina, often used as an "active" (as opposed to inert) support, is an active catalyst for the dehydration of alcohols.

Such facts are the stock in trade of the catalyst formulator and may prove useful if the reader is faced with formulating his own catalyst. The many techniques of depositing catalysts on carriers are beyond the scope of this book but the reader should keep in mind that many of the supports contain relatively small pores (10–100 Å) that can be plugged up during catalyst preparation unless care is used.

Large pores, e.g., 5000 Å, can be introduced into catalysts by adding organic material that is later burned out. For example, 5% of starch or intensively shreaded α-cellulose can be mulled into a finely divided support with or without the active catalyst. The mixture is pelleted or extruded, dried, and oxidized to produce a catalyst with a bimodal pore size distribution. The large pores from the organic additive and the normal smaller pores in the support.

From this discussion, two major factors dominate the selection of a support. (1) It must provide access to the active catalytic material. (2) It must be stable under the conditions of use and regeneration (*4*).

While one ordinarily thinks of carriers in terms of supports for solid catalysts, liquids can be supported too. The vanadium pentoxide catalyst for oxidizing SO_2 to SO_3 is liquid under operating conditions. It may be preferable to support and adsorb some of the highly active homogeneous catalysts being developed rather than to go to the expense of separating them from the reaction product for reuse or to recover the valuable component. For example, the homogeneous catalyst, monochloro(carbonyl)bis(triphenylphosphine)rhodium(I) dissolved in butyl benzyl phthalate has been studied for the hydroformylation of propylene with the catalyst solution supported on silica gel (*7*).

Formed carriers, whether formed by the user or by the manufacturer, may be anisotropic. If the original particles are anisotropic, e.g., platelets or needles, the forming operation can orient them in the direction of flow. For example, tabletted or extruded catalysts made from a material that crystalizes in plates may have a "skin" on them. In some cases, this can cause the reaction rate to be diffusion limited when otherwise it need not be.

A somewhat similar effect can result from die lubricants, especially in tabletted catalysts. One to two percent of flake graphite or, more rarely, talc may be added to the powdered catalyst before tabletting. This lubricates the dies and tablets and makes them easier to eject. At the same time, a shiney coating of plates covers the cylindrical part of the tablet. This coating may or may not cause a diffusion-limited reaction rate, but if the reaction rate is diffusion limited, the coating should be suspect. If practical, organic die lubricants that can be washed or burned out are preferred, e.g., starch, naphthalene, or stearin.

I. Active Catalyst Carriers

It is not possible to draw a definitive line between the active and the inert supports. The author has done this on the assumption that carriers with large surface areas are at least potentially "active."

A. Silica Gel

Silica gel can be prepared with a range of properties, primarily with variations in surface area and pore size. While it has been proposed as a catalyst for a number of reactions, it is not commonly used commercially as a catalyst in its own right. It is, however, rather commonly used as a catalyst support and particularly for vanadia oxidation catalysts, oxychlorination catalysts, and precious metal hydrogenation catalysts.

In the event one wants to composite his own catalysts and does not want to take the risk of plugging up the pores in silica gel with the material to be supported, then the supplier, on special request, will supply purified silica hydrogels. These can be composited with the catalytic material and then dried to develop the internal structure.

One can prepare his own hydrogel from commercial water glass. Acidifying the diluted solution to form a silica hydrosol and allowing this to set to an all-imbibing gel gives good results. The hydrogel holds on to Na^+ rather tenaciously, so exhaustive washing is necessary if the Na^+ ions have any undesirable effects. (Residual Na^+ tends to cause silica-gel based catalysts to sinter if used or regenerated at high temperatures, e.g., $>500°C$).

Silica hydrogels can also be prepared from commercial colloidal silica (See Section I–B).

1. *Supplier of Silica Gels*

Davison Chemical Division of W. R. Grace & Co.

- RD–Gel: as granules. The ABD is 0.7, the surface area 800 m^2gm^{-1}, and the pore volume 0.45 cm^3gm^{-1}.
- ID–Gel: as granules. The ABD is 0.45, the surface area 350 m^2gm^{-1}, and the pore volume 1.04 cm^3gm^{-1}.
- MS–Gel: as microspheres. The ABD is 0.45, the surface area 600 m^2gm^{-1}, and the pore volume 1.00 cm^3gm^{-1}.
- MS–ID–Gel: as microspheres. The ABD is 0.40, the surface area 350 m^2gm^{-1}, and the pore volume 1.80 cm^3gm^{-1}.
- Grade 02: as granules. The ABD is 0.7, the surface area 525 m^2gm^{-1}, and the pore volume 0.30 cm^3gm^{-1}.
- Grade 57: as granules. Large pore volume.
- Grade 81: as powder. The ABD is 0.6, the surface area 600 m^2gm^{-1}, and the pore volume 0.70 cm^3gm^{-1}.

B. Colloidal Silica

Colloidal silica is useful in formulating catalysts as a support, component, or binder. The individual particles are spherical and contain no internal surface. The external surface is covered with SiOH groups and the colloid is usually stabilized with NaOH to a pH of 9 but can be stabilized with NH_3.

Reducing the pH from 9 causes the colloidal silica to gel. The higher the concentration of the sol, the tighter the gel and the smaller the pores. The addition of salt also causes the sol to gel, but here the pore structure is more open, i.e., larger pores. Alcohol also gives an open gel. F^- ions at low pH favors forming —Si—O—Si— from —SiOH.

It is possible to form compounds (metal ions) with the SiOH surface. It is also possible to deposit Ag or Cu and possibly other metals by adding Ag^+ or Cu^{2+} and a suitable reducing agent.

Including a little Al^{3+} in the structure, replacing some of the Si, puts a negative charge on the sol that is independent of the pH. The gelling characteristics are changed so that it is stable in the pH range of 3.5–8. It can be forced to gel by large dosages of salt, by alcohols, ethylene diamine, ethyl lactate, or by evaporating the water.

1. *Supplier of Colloidal Silica*

E. I. Du Pont de Nemours & Co., Inc.

Ludox TM: 50% SiO_2, 0.21% Na_2O; particle size about 22–25 mμ, and surface area (dried) 140 m^2gm^{-1}.

Ludox HS (High Sodium): 40% SiO_2 (0.43% Na_2O) or 30% SiO_2 (0.32% Na_2O), particle diameter about 13–14 mμ, and surface area (dried) 220–235 m^2gm^{-1}.

Ludox LS (Low Sodium): 30% SiO_2, 0.1% Na_2O; particle diameter about 15–16 mμ, and surface area 195–215 m^2gm^{-1}.

Ludox AR (Ammonia Stabilized): 30% SiO_2, 0.25% NH_3; particle diameter about 13–14 mμ, and surface area (dried) 220–235 m^2gm^{-1}.

Ludox AM (Alumina Modified, 0.2% Al_2O_3), 30% SiO_2, 0.13% Na_2O; particle diameter about 13–14 mμ, and surface area (dried) 220–235 m^2gm^{-1}.

Ludox SM: 15% SiO_2, 0.1% Na_2O; particle diameter about 7–8 mμ, and surface area (dried) 350–400 m^2gm^{-1}.

C. Diatomaceous Silica, Diatomaceous Earth, Diatomite, or Kieselguhr

All these names are used to describe the same product. Diatomaceous silica is one of the catalyst formulator's more valuable support materials. It can be a powerful tool in the hands of one who understands it and knows its characteristics, especially since these characteristics may vary from deposit to deposit, with processing conditions, and with supplier.

There are many types of diatoms, so there are many types of diatom skeletons that make up diatomaceous silica. As deposited, they may be mixed with small proportions of clay and sand, but the supplier removes most of these. The iron and alumina that remain are believed to be an essential part of the skeleton.

Natural diatomaceous silica is characterized by having moderate surface area, 10–80 m^2gm^{-1}, and relatively large pores, i.e., 100 Å or more in diameter. The silica surface apparently has some SiOH groups on it for it will dissolve in a hot 5% solution of Na_2CO_3 or react with a slaked lime slurry. One can hardly consider the surface inert. For example, diatomaceous silica is used as a "support" in making solid phosphoric acid polymerization catalysts. However, there is evidence for chemical reaction between the SiOH groups and the phosphoric acid.

Compared to supports such as "activated" alumina, it is relatively non-adsorptive. It is especially useful for formulating catalysts where one may be concerned with getting a reaction product desorbed.

When one heats diatomaceous silica, SiOH groups are converted into Si—O—Si as one would expect. In addition, there is a certain amount of sintering with the disappearance of small pores leaving a larger average pore diameter. During calcination in air, the iron is oxidized so that the original slightly grey earth becomes pink.

When one calcines diatomaceous silica with a small proportion (3–6%) of Na_2CO_3, still more sintering and some fluxing occurs. The net result is still larger pores. Further, the iron in the earth is involved in some way so that it does not get oxidized even in the presence of air. These products are white—whiter than the original silica.

It will be recalled that diatomaceous silicas are mainly used commercially as filter aids. They trap particles in the solution to be filtered and prevent them from passing or "blinding" the filter septum. For example, a rotary filter is given a coat of filter aid from a slurry. Then the solution to be filtered (containing suspended particles) is started through. As the outside of the filter aide coating becomes filled with fine suspended material, it is sliced off with a special knife blade so that a fresh surface of filter aide is exposed. This continues until the filter aide coat is gone and must be replaced.

In a commercial operation, the rate of these filtrations is important. Usually the maximum practical rate is desired. At the same time, a clear filtrate (free of suspended solids) is wanted. The filter aides with the largest pores have the highest flow (filter) rates. These are the white (sodium carbonate calcined) diatomaceous silicas. However, the pores may be so large that suspended material gets through the filter and the product lacks clarity, i.e., is cloudy. Various grades of diatomaceous silica with varying pore sizes are then tried until the maximum flow rate for the desired clarity is obtained.

From this, the catalyst chemist will see that he has a whole spectrum of pore sizes for his catalyst support at his command. It is not enough to try one diatomaceous silica at random and give up when it does not work. If one of them works, then the preceding background information will be invaluable in selecting the optimum support.

1. *Suppliers of Diatomaceous Silica*

Grefco, Inc. supplies many grades of diatomaceous silica under the trade name Dicalite. The name is usually followed by a number or other designation indicating the grade. All are powders.

Johns-Manville supplies many grades of diatomaceous silica under the generic trade name Celite. Usually, a number follows the name, and this indicates the grade. Most are powders but Celite 408 is a ⅛ inch pellet and Celite 410 is a ¼ inch pellet.

If one uses the terms Celite or Dicalite, diatomaceous silica or kieselguhr, it will be clear that these are not definitive descriptions and do not characterize the product.

D. Alumina Hydrates and Active Aluminas

To the chemist encountering it for the first time, the series of alumina hydrates and active aluminas is bewildering. The author has selected those of interest to the catalyst chemist and has tried to make it less bewildering. It must be kept in mind, however, that there are many ways of preparing alumina and so far as the author knows, the physical methods of characterizing catalysts such as X-ray diffraction, surface area, pore volume, etc., do not define their catalytic activity.

Aluminas are probably the most versatile of the catalyst supports and the author would advise the chemist seeking a support to try one of the active aluminas first.

1. *α-Alumina Trihydrate*

This trihydrate is the principal constituent of the North and South American bauxites and is the product of the Bayer process for the purification of bauxites to remove iron, silica, and other impurities. Broadly, the Bayer process dissolves the alumina hydrate in NaOH to form sodium aluminate. The insoluble impurities are separated and the aluminate hydrolyzed to give the hydrate:

$$2\,NaAlO_2 + 4\,H_2O = Al_2O_3 \cdot 3\,H_2O + 2\,NaOH$$

A similar hydrolysis has been used to deposit an adhering coating of the trihydrate on stainless steel wire mesh. The hydrolysis was done at 60°–80°C and the alumina replaced by an excess of aluminum metal (*5*):

$$2\,Al + 2\,NaOH + 2\,H_2O = 2\,NaAlO_2 + 3\,H_2$$

The composite was used as a support for oxidation catalysts for automobile mufflers, but the alumina support can be used for any other type of catalyst where the low pressure drop provided by the wire mesh is needed.

α-Alumina trihydrate loses water slowly at 250°C and more rapidly at 400°–500°C, forming γ-alumina containing about 0.4–0.5% bound water. This is one of the more common, if not the most common, active (or activated) aluminas. Depending on particle size, heating rate, in vacuum or not, the surface area of the γ-alumina is usually in the 250–350 m^2gm^{-1} range. This is the catalytically interesting material. Heating to higher temperatures, either during use or during regeneration, causes the alumina to lose more water and ultimately results in the formation of the inactive α-alumina with the concurrent loss of almost all of the surface area.

2. β-*Alumina Trihydrate*

This trihydrate is formed by CO_2 reacting with sodium aluminate. Made in this way, it is relatively dense. Made by reacting water with amalgamated aluminum metal, it is a relatively light powder with considerable surface area. It has a greater stability towards transformation on heating than the dense form.

The rapid heating of the β-alumina trihydrate in moist air gives γ-alumina. The slow heating (<1°C min^{-1}) of small particles ($<10\ \mu$) in dry air gives η-alumina (*6, 8*). This is another of the interesting active aluminas. It, too, on heating to high temperatures either during use or regeneration, goes to α-alumina and loses its activity.

3. α-*Alumina Monohydrate*

This hydrate is the principal constituent of European bauxite. It is formed as a gelatinous precipitate from aluminum salt solutions by alkali or ammonia. Heating the product above about 400°C decomposes the monohydrate, giving γ-alumina. Both the α- and β-alumina trihydrates can give the α-monohydrate as an intermediate *en route* to γ-alumina.

The several alumina hydrates have other names. α-Alumina trihydrate is also called Gibbsite or Hydrargillite. β-Alumina trihydrate is also called Bayerite, and the monohydrate is called Böhmite.

4. *Suppliers of Active Aluminas*

Aluminum Company of America

Grade F–20: a γ-alumina, 92% Al_2O_3, 6.8% H_2O, 0.8% Na_2O, as powder. The ABD is 1 and the surface area 210 m^2gm^{-1}. Made by heating the α-trihydrate.

Grade H–51: 85% Al_2O_3, 6.0% H_2O, 5.8% SiO_2, 2.0% Na_2O, as microspheres. The ABD is 0.85, the surface area 350 m^2gm^{-1} and the pore volume 0.5 cm^3gm^{-1}. Made from gelatinous α-alumina monohydrate.

C–31 and C–33 α-alumina trihydrate: 64.9% Al_2O_3, 34.7% combined H_2O, 0.35% Na_2O, as powders. C–31 contains 0.01% Fe_2O_3; C–33 0.003% Fe_2O_3. The ABD is 0.85 and the surface area 0.1–0.2 m^2gm^{-1}.

Filtrol Corporation

Grade 86: 7% volatile, 0.07% Fe_2O_3 (volatile-free basis), 0.02% Na_2O and 3.5% SO_3, as 1/8 and 3/16 inch extrudates. The ABD is 0.7 and the surface area 300–325 m^2gm^{-1}.

Grade 90: 20–26% volatile, 0.05% Fe_2O_3 (volatile-free basis), 0.02% Na_2O, 3.5% SO_3, as microspheres. The surface area is 300 m^2gm^{-1}.

Harshaw Chemical Company

Al–0104 T: activated alumina, 99% Al_2O_3, as 1/8, 5/32, 3/16, and 1/4 inch tablets. The ABD is 0.7–1.0, surface area 80–100 m^2gm^{-1}, and the pore volume 0.28–0.33 cm^3gm^{-1}.

Al–1401 P: active alumina, 97% Al_2O_3, as microspheres.

Al–1404 T: active alumina, 96% Al_2O_3, as 1/8, 5/32, 3/16, 1/4, and 3/8 inch tablets. The ABD is 0.75–0.9, the surface area 180–200 m^2gm^{-1}, and the pore volume 0.42–0.51 cm^3gm^{-1}.

Al–1602 T: active silicated alumina, 91% Al_2O_3 and 6% SiO_2, as 1/8 inch tablets. The ABD is 0.85, the surface area 210–240 m^2gm^{-1}, and the pore volume 0.48 cm^3gm^{-1}.

Al–1605 P: active silicated alumina, 92% Al_2O_3 and 6% SiO_2, as powder. The ABD is 1 and the surface area 210–240 m^2gm^{-1}.

Al–1706 E: active alumina, 97% Al_2O_3, as 1/8 inch extrudates. The ABD is 0.5, the surface area 218 m^2gm^{-1}, and the pore volume 0.77 cm^3gm^{-1}.

Al–2810 P: active alumina, 98% Al_2O_3, $<$0.05% Na_2O, 20% volatile, as powder. The ABD is 0.45, the surface area 340 m^2gm^{-1}, and the pore volume 0.76 cm^3gm^{-1}.

Al–3404 P: active alumina, 99+% Al_2O_3, $<$0.008% Na_2O, 10% volatile, as powder. The ABD is 1.0, the surface area 175–200 m^2gm^{-1}, and the pore volume 0.40 cm^3gm^{-1}.

Al–3428 T: active alumina, 98% Al_2O_3, $<$0.008% Na_2O, as tablets.

The ABD is 1.0, the surface area 175 m^2gm^{-1}, and the pore volume 0.49 cm^3gm^{-1}.

Houdry Process and Chemical Company

Grade 100: 98.5% Al_2O_3, 0.4–0.5% Na_2O, as $\frac{3}{32}$, $\frac{1}{8}$, $\frac{5}{32}$, and $\frac{3}{16}$ inch extrudates. The ABD is 0.78–0.82 and the surface area 75–85 m^2gm^{-1}.

Grade 100S: 98.5% Al_2O_3, 0.1–0.2% Na_2O, as $\frac{3}{32}$, $\frac{1}{8}$, $\frac{5}{32}$, and $\frac{3}{16}$ inch extrudates. The ABD is 0.78–0.82 and the surface area 75–85 m^2gm^{-1}.

Grade 200: 90.5% Al_2O_3, 7–8% bound H_2O, 0.4–0.5% Na_2O, as $\frac{3}{32}$, $\frac{1}{8}$, $\frac{5}{32}$, and $\frac{3}{16}$ inch extrudates. The ABD is 0.82–0.86 and the surface area 180–210 m^2gm^{-1}.

Grade 200S: 98% Al_2O_3, 0.1–0.2% Na_2O, as $\frac{3}{32}$, $\frac{1}{8}$, $\frac{5}{32}$, and $\frac{3}{16}$ inch extrudates. The ABD is 0.77–0.81 and surface area 150–180 m^2gm^{-1}.

Minerals and Chemicals Division, Englehard Minerals and Chemicals Corporation

Porocel Low Iron: 81.5% Al_2O_3, 3.0% Fe_2O_3, 10.5% SiO_2, 3.5% TiO_2, as granules. The ABD is 0.9 and the surface area 185–205 m^2gm^{-1}.

Porocel Low Silica (S): 92% Al_2O_3, 2.5% Fe_2O_3, 2.2% SiO_2, 2.8% TiO_2, as granules. The ABD is 0.9 and the surface area 200–220 m^2gm^{-1}.

Porocel Low Silica (B): 88% Al_2O_3, 5% Fe_2O_3, 2.6% SiO_2, 2.8% TiO_2, as granules. The ABD is 0.9 and the surface area 195–210 m^2gm^{-1}.

E. Colloidal α-Alumina Monohydrate

Colloidal or dispersible alumina is useful in formulating catalysts as a support, component, or binder. An α-alumina monohydrate (AlOOH) is available which is dispersible in water and other solvents containing OH groups. It is made up of solid (nonporous) fibers about 100 mμ long, with a rectangular cross section of 5–6 mμ. It has a surface area of 275 m^2gm^{-1} which is rich in AlOH groups. It retains its surface area to higher temperatures than any of the other high surface area aluminas. For example, after heating to 750°C, the area is reduced from 275 to about 240 m^2gm^{-1}. The heating converts it to γ-alumina.

As might be expected from the fiber shape, it forms straw-stack or brush-pile solid structures. The "pores" in this structure are 50–70 Å in diameter and the pore volume is 0.5–0.6 cm^3gm^{-1}. The following characteristics are worth noting.

(1) The particles carry a positive charge.

(2) It will disperse in other liquids.

(3) It is film forming. This can be useful in depositing a film of catalytically active material on an inert smooth support.

(4) It is a good binder.

(5) Low concentrations flocculate clay; higher concentrations disperse clay.

(6) It acts as a base with acids.

(7) It forms chelates, e.g., with benzophenone.

1. *Supplier of Colloidal α-Alumina Monohydrate*

E. I. du Pont de Nemours & Company, Inc.
Baymal: a paste of 84% AlOOH + 12.5% acetic acid.

F. Molecular Sieve Zeolites

Molecular sieves are crystalline aluminosilicates belonging to a class of compounds known as zeolites. Zeolite minerals have been known for about 200 years, but it was not until about 100 years after the discovery of the first zeolites that their ion-exchange properties were recognized. In the 1920's, the first adsorption studies were carried out. These revealed the ability of certain zeolites to accept or reject molecules based on their size. Since that time, about sixty zeolites have been synthesized, some of which are related to known mineral zeolites while others have no known mineral counterparts.

The useful properties of the zeolites stem from their crystal structure, which is characterized by (*a*) a three-dimensional lattice of SiO_4 and AlO_4^- tetrahedra joined by a mutual sharing of oxygen atoms at their apexes to form a network of uniform cavities available only to molecules small enough to diffuse through the pores, (*b*) a highly polar surface created by strong fields, such as those associated with the cations present in the lattice. The various crystal forms in which the zeolites can be synthesized and their various cation modifications can differ considerably from each other in their behavior as adsorbents and as catalysts. There are summaries of the interesting crystal structures of the zeolites (*1*, *2*), as well as a summary of the uses of molecular sieve zeolites (*3*).

The most important commercially available molecular sieve zeolites are Linde types A, X, and Y, as well as mordenite. Zeolite type A has the lowest silica content and the highest cation density of the class and has an effective pore size of 3, 4, or 5 Å, depending on whether the contained cation is potassium (type 3A), sodium (type 4A), or calcium (type 5A). These types are used primarily as adsorbents. The type A has no known mineral counterpart. Types X and Y are structurally related to the rare mineral, faujasite, which has a complex chemical composition in terms of the contained cations (Na^+, K^+, Ca^{2+}, Mg^{2+}). Types X and Y have common characteristics as well as important differences. Type Y has a higher silica content than type X and is structurally more stable. As in the case of type A, both have an ultimate crystal size of about 1–3 μ; the pore size is between 8 and 9 Å, depending on whether the contained cation is calcium (type 10X) or sodium (types 13X and type Y). The large pore size of these materials makes them particularly useful as catalysts. Mordenite has a higher silica content and a lower cation density than any of the other commercial zeolites and can be synthesized with a pore size of 6–7 Å. Unlike the three-dimensional pore structure of types A, X, and Y, however, the pore structure of mordenite is unidimensional; the pore system is analogous to a bundle of tubes which traverse the crystal in one direction. The synthetic mordenite is structurally related to the mineral mordenite.

The catalytic properties of the zeolites are greatly influenced by the silica to alumina ratio, the kinds of cations present in the crystal lattice, and the number of cation deficient sites. The important use of the molecular sieves in cracking catalysts and hydrocracking catalysts has already been described. Replacing the cations in the sieves as synthesized with catalytically active ions, which may or may not be reduced to catalytically active metals, offers an almost limitless opportunity for developing catalysts.

1. *Suppliers of Molecular Sieves*

Chemical Process Division, Norton Company

Zeolon–Na: the sodium form of synthetic mordenite, as powder, 1/16 or 1/8 inch extrudates. The powder ABD is 0.4 and the surface area 440 m^2gm^{-1}. The extrudated ABD is 0.65 and the surface area 400 m^2gm^{-1}. Maximum use temperature is 420°C.

Zeolon–H: the hydrogen or acid form of mordenite, as powder and 1/16 or 1/8 inch extrudates. The ABD of the powder is 0.55 and the surface area 500 m^2gm^{-1}. The ABD of the extrudates is 0.65 and the surface area 450 m^2gm^{-1}. Maximum use temperature is 800°C.

Linde Division, Union Carbide Corporation supplies types A, X and Y in the compositions given above, as powders.

II. Inert Carriers and Supports

The inert carriers are relatively low in surface area and it is usually presumed that, unlike the active carriers, there is no chemical interaction between carrier and the supported catalyst. Most of them have a relatively small surface area, so it is the geometric surface that is important. Porosities are measured in percent rather than cm^3gm^{-1}. They are products of the refractory industries and are stable and usable at high temperatures.

The term "support" can have a different meaning with these inert materials. A fixed bed of catalyst must be held in place in some way. In the laboratory it can be "supported" on a plug of glass wool or on a sintered glass disc. In the pilot plant, a perforated metal disc serves as the "support." In a commercial unit, a perforated metal plate is used as a primary platform but a layer of graded sizes (largest sizes next to the platform and smallest sizes on top) of inert refractory granules is usually placed on the platform and this "supports" the catalyst. This technique prevents catalyst granules, tablets, or extrudates from plugging the holes in the perforated plate and causing excessive pressure drop.

A. Suppliers of Carriers and Supports

Aluminum Company of America

- A-1: 99% Al_2O_3, as powder. The ABD is 0.9, the surface area 20–70 m^2gm^{-1}, and the pore volume about 0.25 cm^3gm^{-1}.
- A-2: 99% Al_2O_3 (90+% α), as powder. The ABD is 0.85, the surface area 0.4 m^2gm^{-1}, and the pore volume 0.25 cm^3gm^{-1}.
- A-3: 99% Al_2O_3 (80% α) as powder. The ABD is 0.9.
- A-5: 99% Al_2O_3 (90+% α), as powder. The ABD is 0.75, the surface area 0.4 m^2gm^{-1}, and the pore volume 0.26 cm^3gm^{-1}.
- A-10: 99.5% Al_2O_3 (95+% α), as powder. The ABD is 1, the surface area 0.1 m^2gm^{-1}, and the pore volume 0.12 cm^3gm^{-1}.
- A-14: 99.6% Al_2O_3 (95+% α), as powder. The ABD is 1.
- T-61: 100% α-alumina as granules from 28 mesh to ½ inch or imperfect cylinders 1 inch in diameter by ⅝ inch long. The ABD is 2, the surface area 0.04 m^2gm^{-1}, and the pore volume 0.015 cm^3gm^{-1}.
- T-71: 100% α-alumina in the same sizes. The ABD is 1.2, the surface area 0.5 m^2gm^{-1}, and the pore volume 0.1–0.2 cm^3gm^{-1}.
- T-160: α-alumina as imperfect cylinders, ½ to ⅞ inches.
- T-162: α-alumina as spheres ¼, ⅜, ½, and ¾ inch diameters.
- T-164: α-alumina as smooth surface spheres ¼ and ⅜ inch diameters.

The Carborundum Company supplies their products in shapes that include spheres, cylinders, granules, and powders. Spheres range in size from 3/16 to ⅜ inch in diameter, cylinders from 3/16 × 3/16 to ¾ × ¾ inches, and granules from 12 mesh to ½ inch.

- α-Alumina: in ten different grades, with surface areas from 0.079 to 0.27 m^2gm^{-1} and porosities from 26 to 48%. The ABD is 1–1.4. In addition, there are five grades of fluid α-alumina powder ranging from 0.6 to 4 m^2gm^{-1} in surface area and in pore volume, from 0.18 to 0.32 cm^3gm^{-1}.
- Silicon carbide: in ten different grades with surface areas ranging from 0.138 to 0.569 m^2gm^{-1} and porosities from 30 to 58%. The ABD ranges from 0.7 to 1.

Mullite (a fused aluminum silicate) in three grades with surface areas of 0.172 to 0.188 m^2gm^{-1}, porosities from 33 to 46%, and ABD's from 0.7 to 1.1.

Fused silica: in ten grades with surface areas from 0.140 to 5.64 m^2gm^{-1}, porosities from 39 to 60%, and ABD's from 0.55 to 1.3.

Fused zirconium silicate: in two grades with surface areas of 0.090 and 0.310 m^2gm^{-1}, porosities of 37%, and ABD's of 1.25 and 1.45.

Fused zirconia: CaO stabilized; surface area is 0.163 m^2gm^{-1}, porosity 43%, and ABD 1.7.

Fiberfrax: alumina–silica ceramic fibers that can be used as a low pressure drop catalyst carrier at temperatures up to 1250°C or as a high temperature filter to trap valuable catalyst dust particles, e.g., Pt, in the exit gas from an ammonia oxidation plant.

Chemical Process Division, Norton Company supplies their inert supports as microspheres and powders, granules, extruded cylinders, spheres, and rings.

Alundum α-alumina: is supplied in eighteen different grades, all with surface areas of <1 m^2gm^{-1}. Porosity varies from 31 to 55% and ABD from 0.7 to 1.25.

Corning Glass Works

Cercor: a cellular (corrugated) thin-walled ceramic in which holes form parallel individual triangular tubes running through a fabricated body. The thickness of the body is 2 to 3 inches and diameters of discs go up to 29 inches. They may be used at temperatures up to 1150°C.

T40–75: 360 holes per square inch of frontal area, with the height of the triangular tube opening 0.040 inches and the base 0.075 inches. Web (ceramic) or wall thickness is 0.005 inches. The ABD is 0.5.

T20–38: 1000 holes per square inch of frontal area, with the height of the triangular opening 0.020 inches and the base 0.038 inches. The wall (ceramic) thickness is 0.004 inches and the ABD is 0.065.

E. I. du Pont de Nemours & Co., Inc.

Torvex: ceramic honeycomb made from α-alumina or from mullite. In one form, there are parallel hexagonal tubes that go straight through the honeycomb. Available cell openings are 1/8, 3/16, 1/4, 3/8, and 3/4 inches. In a second form, the tubes slant 45° from the open face of the honeycomb. In a third form, there are half cells angled 90° to each other and 45° to the face of the honeycomb. The slanting half cells intersect half cells slanting in the opposite direction, repeatedly giving cross flow rather than the parallel flow provided by the other two types. Available cell openings are 3/16, 1/4, and 3/8 inch.

All three types are available in 12 inch square blocks from ½ to 2 inches thick.

The α-alumina honeycombs can be used at temperatures to 1500°C and the mullite to 1350°C in air. The ABD varies with the cell size and composition and ranges from 0.16 for ¾ inch cells of alumina to 0.55 for ⅛ inch cells of alumina or mullite.

U. S. Stoneware, Inc.

Body 99: 99% α-alumina as spheres in sizes ranging from ¼ to 2 inches in diameter. The ABD is 1.9.

Body 87: 87% alumina, 12% silica in the same sizes. The ABD is 1.2.

Denstone 57: a ceramic containing 55.5% SiO_2 and 39% Al_2O_3, as spheres from ¼ to 2 inches in diameter (ABD 1.25–1.4) and extruded pellets from ⅛ × ⅛ inches to ⅞ × 1¼ inches (diameter × length). The ABD is 1.4.

Ceratherm 500: a ceramic containing 49% SiO_2 and 45.6% Al_2O_3 as spheres in the same sizes as the above spheres. The ABD is 1.25–1.4.

REFERENCES

1. Barrer, R. M., *Endeavour* **23,** No. 90, 122 (1964).
2. Breck, D. W., *J. Chem. Educ.* **41,** 678 (1964).
3. Collins, J. J., *Chem. Eng. Progr.* **64,** No. 8, 66 (1968).
4. Dowden, D. A., Joint Meeting, Inst. of Chem. Engrs. and Ver. Dutch Ingrs., Brighton, Engl., April 24–26, 1968.
5. Leak, R. J., Brandenburg, J. T., and Behrens, M. D., *Am. Chem. Soc. Div. Petrol. Chem. Preprints* **12,** No. 4, A-43 (1967).
6. Prettre, M., Imelik, B., Blanchin, L., and Petitjean, M., *Angew. Chem.* **65,** 549 (1953).
7. Rony, P. R., *J. Catalysis* **14,** 142 (1969).
8. Tertian, R., and Papée, D., *J. Chim. Phys.* **55,** 341 (1958).

AUTHOR INDEX

Numbers in parentheses are reference numbers and indicate that an author's work is referred to although his name is not cited in the text. Numbers in italics show the page on which the complete reference is listed.

A

B

N

O

P

T

U

V

W

Y

Z

SUBJECT INDEX

A

D

E

F

G

H

I

J

K

L

O

P

T

U

V

W

X

Z